THE OAKWOOD PRESS

# PINFIRE PISTOLS, REVOLVERS AND AMMUNITION HANDBOOK

Steve Jordan

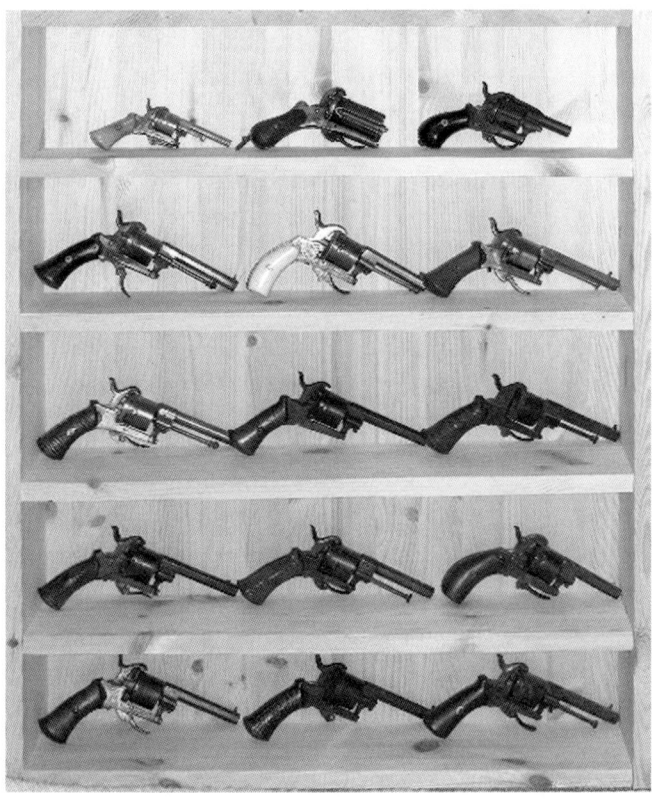

THE OAKWOOD PRESS

© Steve Jordan, 2023
ISBN 978-0-85361-767-9

Published by The Oakwood Press, 2023

Printed by
Claro Print, Office 26, 27, 1 Spiersbridge Way,
Thornliebank, Glasgow G46 8NG

How do you get pinfire ammunition in Hawaii? The *Pacific Commercial Advertiser* for the 27th June, 1868 has the answer, order from Eley's of London.

*Library of Congress*

Published by
The Oakwood Press, 54-58 Mill Square, Catrine, KA5 6RD
Telephone: 01290 551122      Website: www.stenlake.co.uk

# Contents

Introduction .................................................................................. 5

## Chapters

| | | |
|---|---|---|
| One | The Pinfire System of Casimir and Eugene Lefaucheux | 7 |
| Two | Proof Marks Found on Pinfire Revolvers | |
| | Belgium | 29 |
| | France | 30 |
| | Germany | 31 |
| | Great Britain | 31 |
| | Spain | 32 |
| Three | European Pinfire Manufacturers and Patent Holders | |
| | Belgium | 33 |
| | France | 80 |
| | Germany and Austria | 93 |
| | Italy | 102 |
| | Spain | 108 |
| Four | Evolution of the Pinfire Cartridge | 113 |
| Five | European Pinfire Ammunition Manufacturers | |
| | Austria | 121 |
| | Belgium | 122 |
| | England | 125 |
| | Czech | 133 |
| | France | 135 |
| | Germany | 140 |
| | Italy | 144 |
| | Spain | 146 |
| | Unknown | 148 |
| Six | United States of America Pinfire Ammunition Manufacturers | 149 |

## Appendices

| | | |
|---|---|---|
| One | Other Eibar Gunsmiths | 157 |

References and Sources .................................................................. 160

Workers busily filling and finishing cartridges in an American ammunition factory, explosions were common in these places. Over 1.7 million 12mm pinfire cartridges were made in the U.S.A during the Civil War. *Author's collection*

# Introduction

The issue of a patent to Casimir Lefaucheux in 1845 for a breech-loading pistol using a self contained cartridge rendered the cap and ball system obsolete overnight. Dozens of patents over the next ten years smoothed out any imperfections with the original concept and the pinfire revolver quickly became the weapon of choice for many thousands of army officers across Europe who had to provide their own sidearm, as well as civilians who felt the need for a personal protection weapon. Despite the pin cartridge being superseded by the invention of rim and centre fire cartridges in the mid 1840s it still managed to corner much of the European handgun trade with only north America choosing to go with the other ammunition designs. The American Civil War of 1861—1865, and the desperate need for weapons gave the pinfire its chance to reach the mass market in the U.S.A and thousands of revolvers were imported by both the North and South.

In Europe pinfire revolvers were made in their hundreds of thousands in the major arms centres: Liege in Belgium, Paris in France, Brescia in Italy, Solingen in Germany and Eibar in Spain. Strangely, only a very few were made in England, but thousands were imported in the white* and proofed and finished in Birmingham for sale to the home market. Some English patents were issued for special loading systems, but the base weapon was normally an imported Belgian gun.

At its height of popularity the manufacture of pinfire weapons, ammunition and accessories employed many thousands of European foundry workers, grip makers, barrel makers, spring and lock makers, assembly workers, finishers and engravers. At the end of the 19th century the Spanish arms making town of Eibar had 1,100 registered gunsmiths out of a population of only 6,000, most of these were still making pinfire weapons. By 1914 this number had risen to 1,528 gunsmiths working in 40 armouries. The pinfire era began in the 1850s and continued well into the 1930s. The scope for collectors is huge, limited only by space and budget.

When I began collecting pinfires in the early 1990s there was only one English language reference book available, *The Pinfire System* by Smith and Curtis, two American collectors. This book was very well researched and written, but expensive and hard to get hold of in the U.K. Since then there has been a revised and updated edition of this book, titled *Systeme Lefaucheux* by Chris Curtis, but the same drawbacks apply. The ever increasing prices and interest shown in pinfire lots at auction show that the collector base for pinfire weapons is constantly growing, but still

---

* A fully assembled weapon needing final finishing of wood and metal parts.

there is no handy, reasonably priced handbook for collectors. The weapons and ammunition boxes shown and described inside are not all museum quality pieces but are of the type and condition that the average collector would find in any antique shop or auction. With 275 illustrations including over 100 different weapons, 50 patent drawings and details of over 270 gunsmiths and 29 ammunition manufacturers who have profited from the pinfire boom it is hoped that this publication will fill the void.

1880 advertisement for Hippolit Mehles Berlin gun and sporting goods emporium. 'The largest Depot in Germany' with a stock of 2,000 revolvers and prices from 8 to 25 Marks. Like this example most of Hippolit's advertising featured pinfire revolvers of differing types.

*Author's collection*

# Chapter One

## The Pinfire System of Casimir and Eugene Lefaucheux

**1802** At 6am on Wednesday 27th January, baby Casimir was born to Marie and Pierre Lefaucheux in the small village of Bonnetable near Le Mans, France.

An early interest and aptitude in mechanical things, especially firearms saw him apprenticed to the gun trade with a gunsmith near Le Mans.

**1814** At the age of twelve Casimir was again apprenticed, this time with the famous, gunsmith Pauly, at 4 Rue des Trois, Paris. Two years earlier Pauly had invented a breech-loading system for firearms. Shortly after Casimir started at the Paris works Pauly left for an extended trip to England and Henri Roux took over the management.

**1822** Ownership of the Pauly factory was transferred to Eugene Pichereau.

**1827** Casimir succeeded Pichereau as manager and then purchased the works from him. Included in the sale were all patent rights for the Pauly System granted to Roux and Pichereau. On 30th October he married Francoise Constance Faivre, the daughter of a prominent local family.

**1828** Lefaucheux was granted his first patent, 3590 of 10th May, for a Pauly style long arm.

**1832** His son, Eugene, was born on the 14th September.

**1833** He was granted his second patent, 5525, on 28th January. This was for a tip down barrelled shotgun with an improved gas seal. The address of the factory was now given as 5 Rue Jean Jacques Rousseau. A first addition to patent 5525 was granted on 13th March which shows the addition of sliding barrels, further improving the gas seal.

**1834** Two more additions to 5525 were granted on 29th October and 22nd November. These show the factory address as now 10 Rue de la Bourse, Paris.

**1835** The fourth addition to 5525 granted on 31st March shows for the first time a drawing of the pinfire cartridge as we know it now. On 27th June a fifth and final addition to 5525 was granted for a breech-loading pinfire shotgun.

During the period 1832-35 Lefaucheux entered into licensing agreements with several Paris gunsmiths, allowing them to manufacture pinfire weapons for a fee, each gun to be marked 'Lefaucheux Brevette'. Included in the agreement was a clause which gave Casimir sole rights to supply pinfire arms to the French military. In a surprise move in December he sold his business to C. Jube de la Parelle for 25,000 Francs. The sale included all royalties from his previous agreements except for the one with the gunsmith Le Page and the ownership of all patent rights. He also agreed to a non-competition clause which excluded him from making firearms for a period of twelve years. Casimir and his family retired to the village of Pont de Genoa in the Sarthe region, near where he was born. Whilst he was at Pont de Genoa he placed patents for a horse-drawn tractor and an improved cider press.

**1845** Following the expiration of his patents Casimir submitted a new patent for a single shot pistol which was granted on 2nd May as patent 1371.

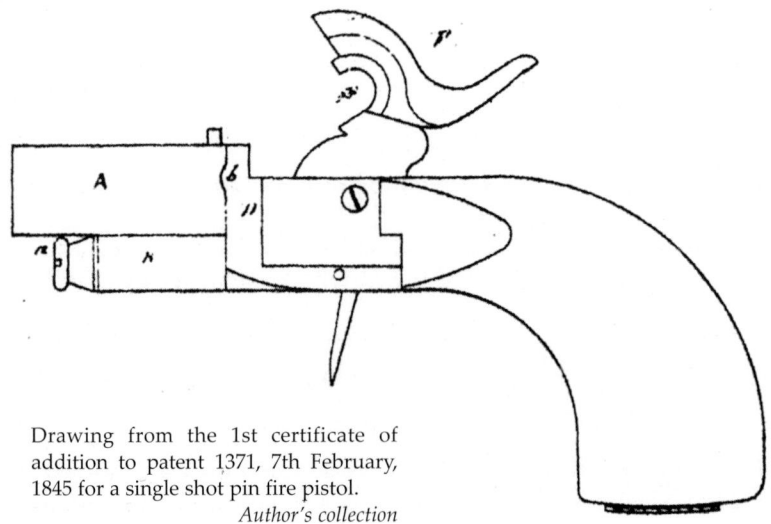

Drawing from the 1st certificate of addition to patent 1371, 7th February, 1845 for a single shot pin fire pistol.
*Author's collection*

## The Pinfire System of Casimir and Eugene Lefaucheux

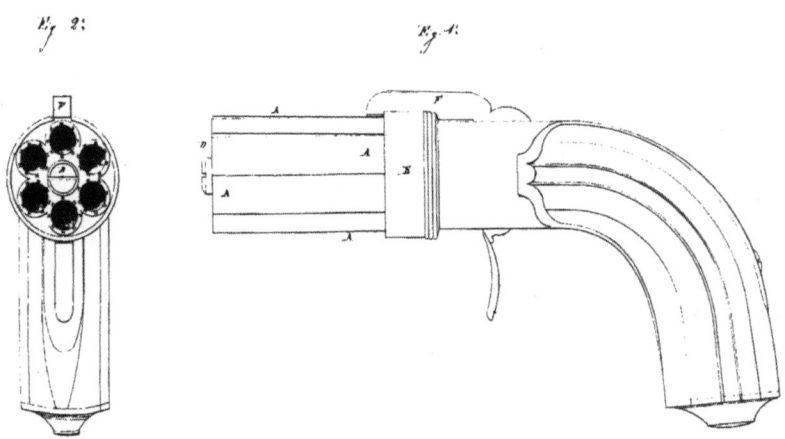

The second drawing from the 1st addition to patent 1371 of 1845 showing a 6 shot, double action revolver with the trigger exposed with no guard and a bar hammer. Removing a retaining screw from the front of the cylinder pin allows the barrel group to slide off for loading, to unload the cylinder pin is used as an extractor. This drawing was probably included only to protect ideas as it does not seem to have ever been built.

*Author's collection*

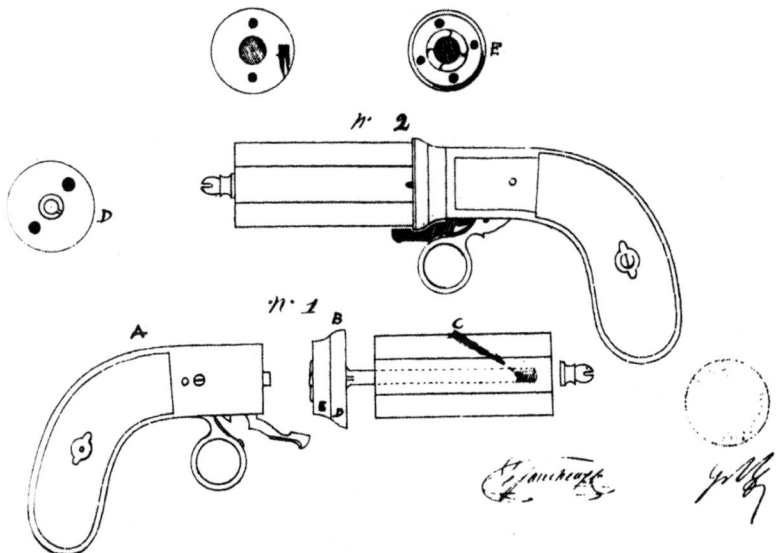

This drawing, from the second certificate of addition to patent 1371, describes a four cylinder, ring triggered revolver very similar to the design by the Belgian gunsmith Mariette. They were also produced in five and six barrel versions. *Author's collection*

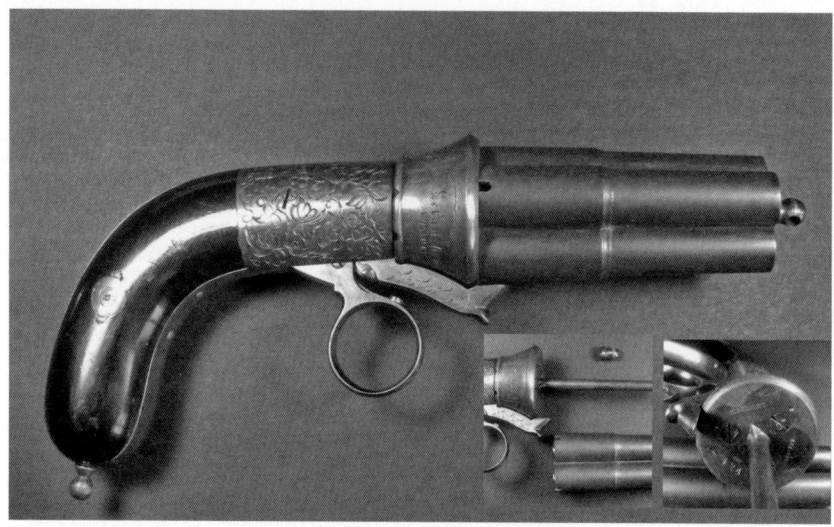

A second certificate to patent 1371 four cylinder underhammer pinfire revolver. First comes the patent, later any minor improvement to the item can be covered by an additional certificate. You can have as many certificates as you want provided that they do not radically change the item which would require the issue of a new patent. The description given exactly describes the weapon shown.

*Author's collection*

Detail from the *London Illustrated News* for 5th July, 1851, it shows Lefaucheux's highly decorated, acid etched Medal of Honour winning revolver. Also shown is a very early pinfire round with a copper base, cardboard tube and a conical lead round.

*aaronnewcomer.com*

**1847** At the end of his 12 year sabbatical from practical gunsmithing Casimir returned to Paris, and re-purchased his works at 10 Rue de la Bourse from Jube de la Perelle. A fourth addition to 1371 was granted on 12th April.

**1848** The 5th, and last certificate of addition to patent 1371 was added on 2nd February.

**1851** Casimir exhibited at the Great Exhibition in the Crystal Palace and was rewarded with a Medal of Honour for his heavily decorated revolving pistol.

**1852** Casimir died on 9th August at the age of 50 and was interred in the Mont Martre Cemetery. Nineteen year old Eugene had been apprenticed to his father from an early age and was well versed with the pinfire system. Leaving Maison Lefaucheux in the charge of his mother Francoise Eugene left for Liege in Belgium, then the centre of European arms manufacturing, to further increase his knowledge of the gun trade

**1853** In July the Governor of Egypt, Said Pasha, purchased three carbines and a pair of premium grade pistols from the shop at 37 Rue du Vivienne which later exhibited at the International Exposition in Paris in 1855. On 14th September Mme Lefaucheux was granted patent number 17391 for a mechanism to eject spent cartridges automatically.

**1854** On his return to Paris Eugene filed the first patent in his own name since taking over the family firm. Dated 15th April and given number 19380 it was granted on 10th June. It described a heavy framed six shot, breech-loading revolver using metallic cartridges and with a distinctly American outline. Eugene had met Samuel Colt and admired his work. Having secured his breech-loading revolver with a British patent, number 955, he then issued a new addition to his French patent with a drawing of his revolver which more closely resembled the production version that we now know as the Model 1854. This included the familiar butt design, a side-loading gate and a captive ejector rod held in place with a leaf spring. This was the first lefaucheux handgun to have E.LEFAUCHEUX stamped on it, to distinguish it from his father's designs.

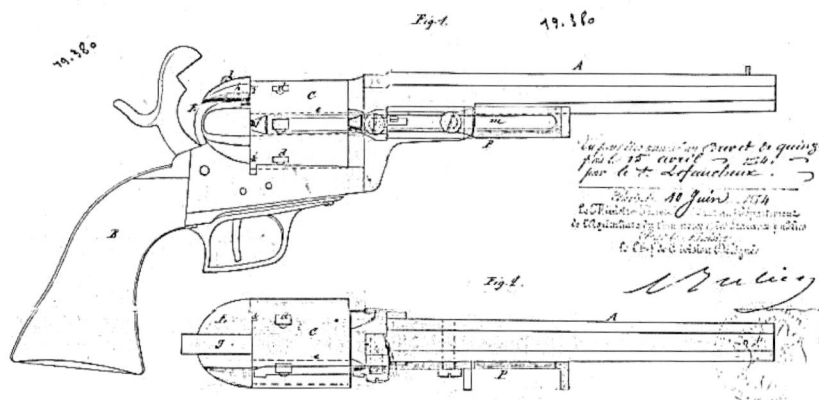

Eugene Lefaucheux's first patent, number 19380 dated 10th June, 1854.   *Author's collection*

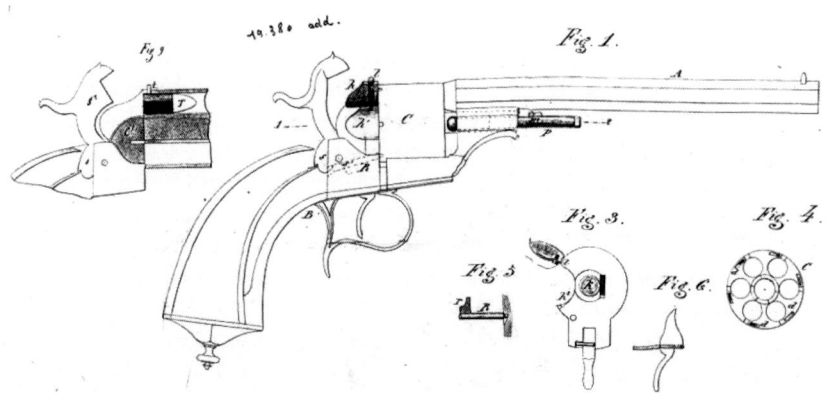

First addition drawing to patent 19380, showing the more familiar layout of the Model 1854. This has the spur added to the trigger guard and the more familiar European-shaped butt.
*Author's collection*

Model 1854's come in three types, firstly the very early ones which faithfully copy the patent drawing with concave recoil shields and octagonal barrel. Then came the mid version with concave recoil shield and a round barrel with an octagonal section next to the cylinder. Lastly, and by far the most common are the late version which has a convex recoil shield and the same part round barrel. The example shown above, serial number 112252, is a late version 12mm Model 1854 produced in the Paris factory. It is marked on the frame beneath the cylinder with the broken gun over LF trade mark and the legend E.LEFAUCHEUX Bte S.G.D.G A PARIS impressed on the top of the barrel.

*Author's collection*

This 9mm example of the Model 1854 was manufactured in Liege under licence by Perlot Freres of 39 Avenue d'Avroy. Rarely made in 9mm the missing metal butt plate and the replacement loading gate can be excused by the scarcity of this calibre weapon.

*Author's collection*

A Liege proofed Model 1854 from the Lefaucheux factory at 12 Quai de Fragnee made between 1860 and 1869. This 12mm, cavalry example has the pre-1893 Liege proof mark and the 1853-77 crowned E inspector's mark. The octagonal barrel section is stamped E.LEFAUCHEUX INVr BREVETTE on the left hand side. The spur on the trigger guard is to provide extra control whist using the weapon mounted on a horse, hence the modern nickname of the 'cavalry' version.
*Author's collection*

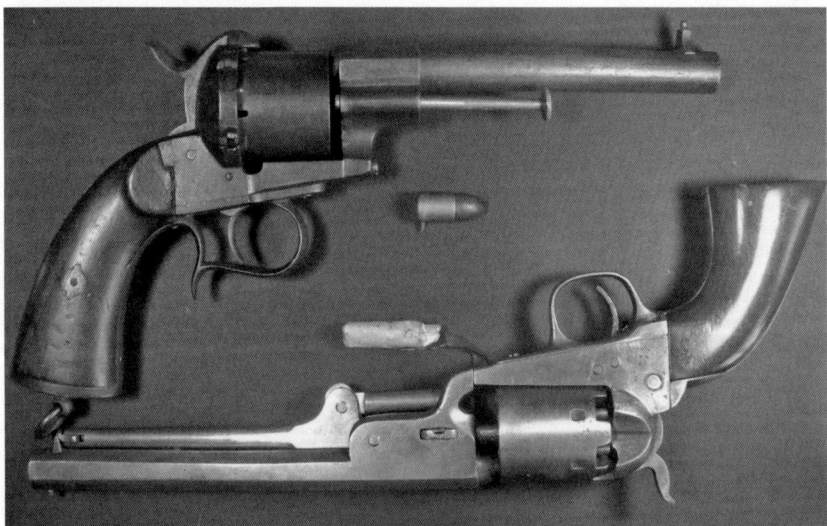

The Model 1854 against its main rival, the Colt 1851 Navy revolver. It takes about 10 seconds to load the paper cartridge into the cylinder, ram it home with the under barrel lever and place a percussion cap on the nipple, but only a couple of seconds to load a pinfire cartridge into the Lefaucheux. In the heat of battle the difference between 60 seconds to load the Colt compared to the 12 seconds to load the M1854 could mean life or death.
*Author's collection*

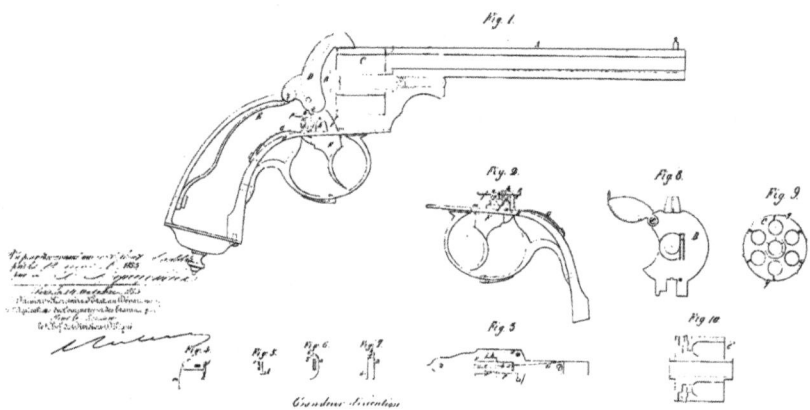

The drawing for the second addition to patent 19380 showing the spurless hammer on this first Lefaucheux double action revolver.
*Author's collection*

9mm version of the second addition to patent 19380. Manufactured in the Paris factory and then exported to Britain where it was proofed at Birmingham. Dating is easy with this weapon as the crowned LF trade mark stamped on the cylinder also has the date 1867. Note the bag grip on this production model instead of the earlier style used on the Model 1854.
*Author's collection*

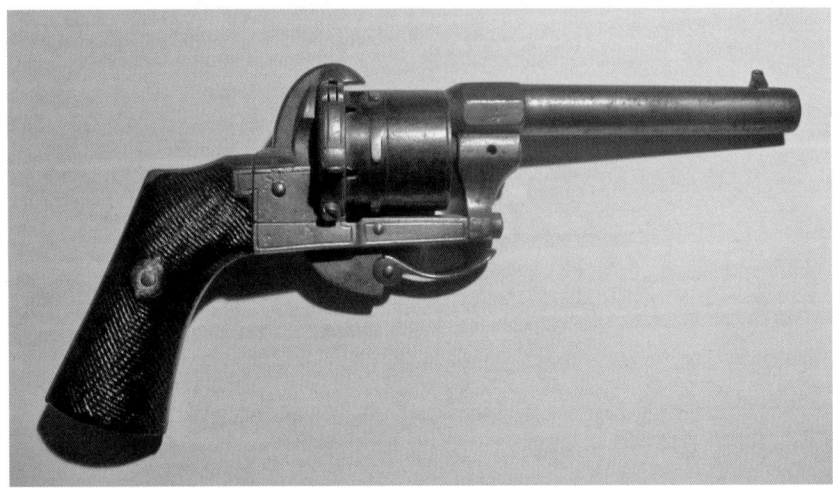

No makers name is visible on this 7mm, Liege proofed copy of this Lefaucheux patent 19380 2nd addition double action revolver, but the quality is there for it to be by Lefaucheux or Pirlot. The post-1862 drop folding trigger is also featured.

*Author's collection*

**1856** Lefaucheux was awarded another patent, number 29055, dated 5th September, for a dual ignition handgun supplied with two cylinders, one pinfire one percussion, the hammer shaped to be able to strike both styles. This double action revolver was designed for the military when on campaign in foreign climes where the availability of pinfire ammunition might be scarce, and the soldier could then utilise the basic cap and ball cylinder.

**1857** On 16th April an advertisement appeared in the *Public Ledger* newspaper advising English gun merchants that the United Belgian Firearms Manufacturers had opened a wholesale depot in London for the sale of licensed Colt, Adams and Lefaucheux revolvers at their Leadenhall Street agency.

Advert from the *Public Ledger* of 16th April, 1857.
*Author's collection*

IMPORTANT to MERCHANTS, SHIP-PERS and GOVERNMENT CONTRACTORS.—The UNITED BELGIAN FIREARMS MANUFAC-TURERS, in consequence of the great demand for their Arms in . England, have established a LONDON DEPOT, and empowered their Agent to take orders at the Manufacturing Prices free in bond London. They are Licencees for Colt's, Adam's and Lefaucheux Re-volvers. Colt's at 50s. each; Adam's, 72s. 6d.; Le-faucheux, 72s. 6d.; Government or Enfield Minié, 45s. 6d. — D. CAHN, Agent, 3, Leadenhall-street, London.

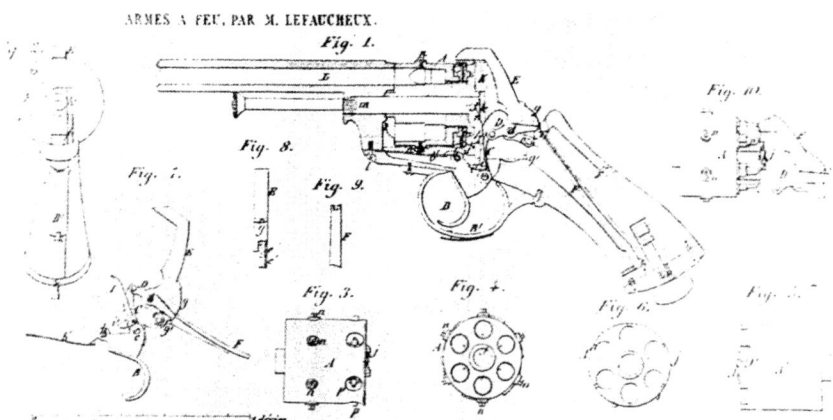

Drawing for patent 29055.  *Author's collection*

**1858** Following exhaustive trials aboard the French Navy vessels *Le Suffren* and *La Bretagne* between the Colt, Adams and Lefaucheux revolvers it was decided that the navy would adopt the 12mm Model 1854 revolver to replace its obsolete percussion small arms. The contract was dated 8th May. It should be noted that for some reason Eugene was unable to produce the weapons fast enough for the navy and even before the contract was signed production had been taken out of his hands, the *Ulsterman* newspaper reporting in April that the French government had ordered that 2,000 revolvers should be made in the Government Arsenal at St Etienne and that Lefaucheux would be indemnified for the infringement of his patent. These Model 1854's made at the arsenal had some minor differences to the standard and were re-designated the Model 1858. A very few of the Lefaucheux-made weapon entered service and do not carry the usual St Etienne markings except for the Navy acceptance mark of an anchor stamped into the butt plate.

Advert from the from the *London Evening Standard* of 20th July, 1858

*Author's collection*

This was a very busy year for Maison Lefaucheux as no sooner had the French Navy been converted to the pinfire system then Queen Isabella of Spain decreed that the replacement for the ageing Beaumont Adams percussion pistols used by the army would be the Model 1854. Later decrees also made it the official side arm for the National Guard, the Fixed Regiment at Ceuta, Morocco, and the regulation private purchase side arm. These weapons were to be made under licence at the Royal Armoury at Trubia by the civilian firm of Orbea Hermanos in Eibar. Arsenal-made weapons are marked TRUBIA and the year of manufacture and those by Orbea with ORBEA HERMANOS-EIBAR.

The Risorgimento, or the Unification of Italy, was also in full swing and Lefeaucheux received orders from the Kingdom of Sardinia to re-equip the navy with Model 1854 revolvers. More orders came in from the warring parties. In 1859 the Piedmontese Government ordered 40,000 revolvers and a year later the Turin authorities ordered another 50,000.

**1860** Already much quicker to load than a percussion revolver Eugene was always looking for ways to improve the speed of loading and ejection of his firearms. To this end he submitted drawings to the Patent Office for revolver with a hinged, swing out cylinder. The

---

### Lefaucheux's Revolvers.

Just received, an invoice of LEFAUCHEUX'S, ADAMS', and COLT'S REVOLVERS.

—Also—

A a fine assortment of Pocket and Saloon PISTOLS, CARBINES, CAPS, etc., etc.

**LION & PINSARD,**

ja31 2ptf        65 and 68 Royal street, corner Bienville.

---

Advertisement from the *New Orleans Daily Crescent* for 3rd February, 1859. Lion and Pinsard were haberdashers (in the United States usage) and could supply almost anything a well-dressed man would need from their premises on Royal Street.   *Library of Congress*

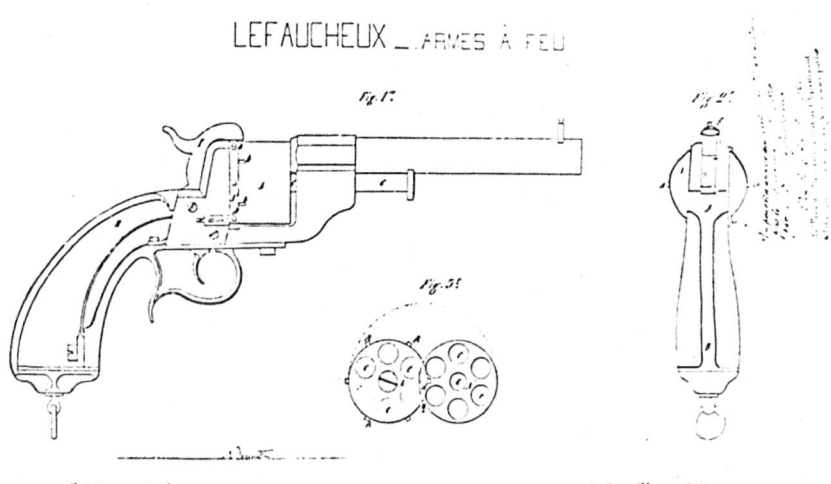

Technical drawing for the second addition to patent 19380 showing the swing out cylinder. Having protected his idea Eugene did not follow through with a production model.

*Author's collection*

cylinder pin was not fixed and pulling it out allowed the cylinder to swing out on its hinge. The cylinder pin was then used to eject the spent rounds. Once reloaded the cylinder was put back into position and the cylinder pin was replaced, locking the parts together. This second certificate of addition was to patent 19380 granted in April.

**1861** On 12th April the sound of artillery resounded around Charleston Harbour as the new Confederate States of America began the bombardment of the Federal-held Fort Sumter. The American Civil War had begun and suddenly thousands of civilians had to be enlisted, trained and armed. To this end agents were quickly despatched to Europe to purchase as many weapons as they could. Confederate major, Caleb Huse left for England on the same day that Sumter was attacked, quickly followed by the Union agent, Colonel George Schuyler. Following his arrival in Liverpool on 10th May Huse made a beeline to London to the works of the London Armoury Co where he purchased 12,000 Enfield rifles, outwitting and out-bidding Schuyler. Having cleared out England, Huse headed for Europe where he continued to purchase top quality weapons for the Confederacy. Having been beaten in England Schuyler proceeded to France where he purchased 10,000

Model 1854 revolvers and 200,000 12mm pinfire cartridges from Lefaucheux, In December another Federal agent, Ordnance Major P.V. Hagner, purchased another 2,000 Model 1854's from Godillot of Paris. By December only 1,500 of the revolvers had reached America and it was assumed the last shipment was lost at sea or captured by a Confederate commerce raider. A few Model 1854's are known to have been sent to the C.S.A, mainly in crates marked 'machine parts' and since most of the records were destroyed during the war we will never know exactly how many.

On 26th March Eugene was awarded his first U.S patent, for a breech-loading pinfire design usable for both rifles and pistols.

In March the Italian Government ordered 5,000 of the Model 1858. This had a shortened barrel and the ejector rod and mount was removed. The poor quality metal used in the Italian-made weapons meant they often bent or broke. This version was named the Pistola a rotazione da Carabinieeri Reali Modello 1861, or Model 1861 Royal Police revolver. These remained in service until 1874.

**1862** The most common item seen on small calibre pistols from this date onwards was patented by Eugene on 2nd December. The drawing for patent number 55784 shows a fairly standard weapon, but the new innovation was a folding trigger This meant that whilst the gun was not in use the trigger could be safely folded up underneath the frame precluding accidental discharges when putting the weapon in a pocket and so on. This simple safety feature saved the cost of a trigger guard and was quickly adopted by the arms trade, especially in Liege.

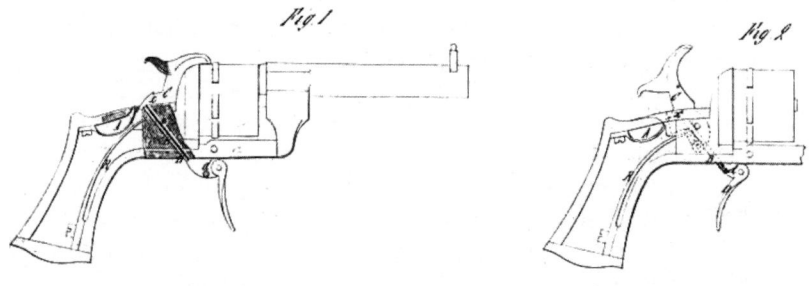

Drawing for patent 55784 showing the soon to be ubiquitous folding trigger.
*Author's collection*

A 9mm version of patent 55784 made at the Lefaucheux factory in Liege. It has fine acanthus leaf engraving on the frame and pin shield as well as carved and cross-hatched grips.
*Author's collection*

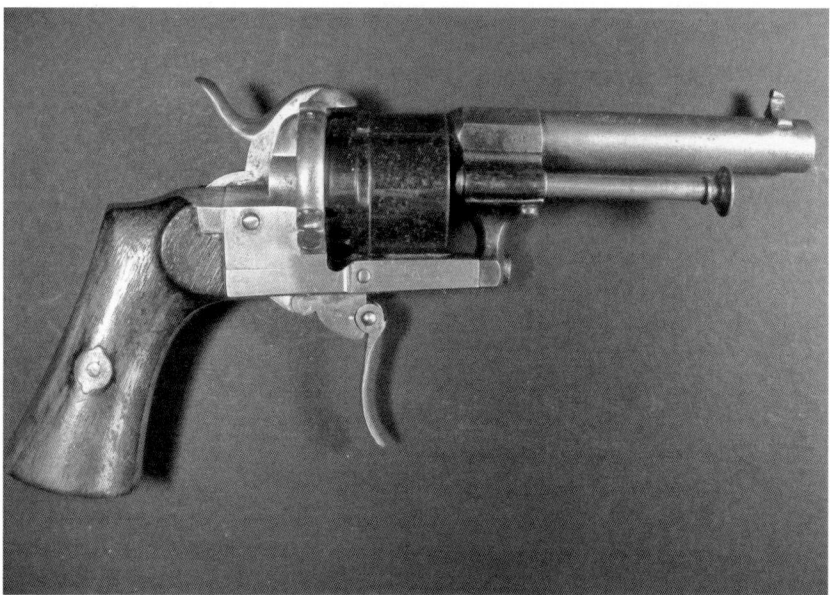

Another fine example of the 1862 patent 55784 revolver, this time in 7mm calibre. Once again it is from the output of the Lefaucheux Quai Fragnee factory in Liege.   *Author's collection*

**1863** Eugene was approached by the Spanish Government and asked to provide a design for a new 12mm, single action side arm to replace the Model 1854 to be produced at the Oviedo Arsenal. The result was a rather plain, but rugged and easy to make, weapon known as the Model 1863. His design was approved on 19th October. Production began in late 1864 and continued until the mid 1870s with approximately 8,500 being made. These revolvers are marked simply with OVIEDO and the year of manufacture punched on to the right hand side of the frame.

Sweden becomes the fourth nation to adopt the Model 1854 revolver for their military with an order for 2,000. They can be identified by the three crowns acceptance mark on the barrel lug.

Norway quickly followed with an order for 1,500 Model 1854's. They had previously ordered a few for testing in 1859, but it took the Norwegian Government another four years, and the order from their neighbour Sweden to decide. Basically the same as the Swedish revolver these were marked with the Norwegian acceptance mark, a rampant lion holding a battleaxe and designated the Model 1864. In 1867 the Norwegian Government decided to manufacture the Model 1864 under licence at their Konigsberg Arsenal: in the end they only made a further 200 weapons. The Model 1864 had a very long life for a military arm,

Model 1863 produced at the Oviedo arsenal in 1869. Lefaucheux design, but not Lefaucheux quality. *Author's collection*

A Lefaucheux-manufactured Norwegian Model 1864/98 fitted with its strengthening top strap, accompanied by a box of regulation ammunition. *Aaronnewcomer.com*

Swedish military Model 1854 with the three leopard head acceptance mark. *gotavan.se*

NOTICE on E. LEFAUCHEUX'S PATENT SYSTEM of SIX-SHOTTED REVOLVING PISTOLS and RIFLES, adopted as the ORDNANCE ARMS of the FRENCH, SPANISH, ITALIAN, DANISH, and other NAVIES.—Special Manufactory of Breech-loading Fire-arms, No. 104, RUE LAFAYETTE, PARIS—Patented by the French Government for his Invention and Improvement in Revolver Fire-arms, an Improvement applicable to Guns of all descriptions, Rifles, and Pistols, Firing Six Charges, and which may advantageously supply the place of all kinds of Pistols and Firearms hitherto known.—These Arms have but one barrel, the balls are forced, and their aim and range are equal to the best firearms hitherto known, whilst the rapidity with which the six charges can be loaded and fired, renders them far superior. For several years past the principle of Revolver Pistols has occupied the attention of military and scientific men. Mr. Colt, in America, and Mr. Adams, in England, have each invented a Revolver, which has met with a certain share of success. His Excellency the Minister of the Imperial Navy of France, having made a selection of all the different systems of Pistols, those of Colt, Adams, and Lefaucheux, were by his order submitted to a series of comparative trials, the result of which was, that on the 16th of September, 1856, the system of Lefaucheux was by the Council of Naval Armaments adopted for the use of the French fleets. The Commission acknowledged that the Pistol invented by Lefaucheux possesses the advantage of being loaded and unloaded with ease and expedition, even during the night; that it is provided with a ramrod for unloading or drawing the charge, which, being introduced into one of the cavities of the cylinder, prevents the Pistol from being cocked, and thus presents every security in carrying this description of Arms; independently of which, the system is so simple, that no other appurtenance than a pocket for the cartridges is required. They are superior to the latter, from the celerity (less than a minute) with which they may be loaded and fired by means of cartridges after the system of Mr. Lefaucheux the father, for which he obtained the medal of honour at the Universal Exhibition of 1855. To those persons who may happen to have exhausted their stock of cartridges the Pistols of Lefaucheux are provided with an iron breech and chimney, by means of which they may be used as an ordinary Pistol, preserving at the same time the advantage over the latter of being loaded or unloaded without requiring any part to be taken to pieces or being obliged to fire. This moveable breech may be used indefinitely with or without cartridges.—SOLE AGENT in LONDON: Mons. A. D. CARIL, 2, JERMYN-STREET, HAYMARKET, S.W.

Advertisement from the *Volunteer Service Gazette* from the 18th March, 1863 edition.
*Author's collection*

remaining in service until 1910. In 1898 they were withdrawn to the Konigsberg Arsenal where they were modified with the addition of a top strap with an integral rear sight, designated the Model 1864/98 and re-issued back to their units.

1864   Eugene entered an application for a further patent on 31st October, granted as number 64960 on 17th December. This patent covered multi-chamber, double action revolvers with two over and under barrels. The double-decker cylinder was bored for eighteen cartridges in two concentric rings, twelve 9mm on the outer ring firing through the top barrel and six 7mm on the inner, firing through the lower. The hammer first strikes the pin of a cartridge on the outer ring then on the next trigger pull the cylinder rotates to allow the hammer to strike a sliding bar which moves vertically to strike the pin of the cartridge on the inner ring and so on. Although the drawing shows a dual calibre cylinder the production models seem to have only been chambered in 7mm.

1865   A first certificate of addition to 64960 was awarded in February for a 20 shot revolver. The drawing shows a cylinder with two concentric rings of ten chambers bored to 7mm. The redesigned hammer strikes in turn the outer ring then the inner ring; again the outer ring fires through the top barrel and the inner through the lower. The use of the fluted cylinder made this weapon very light,

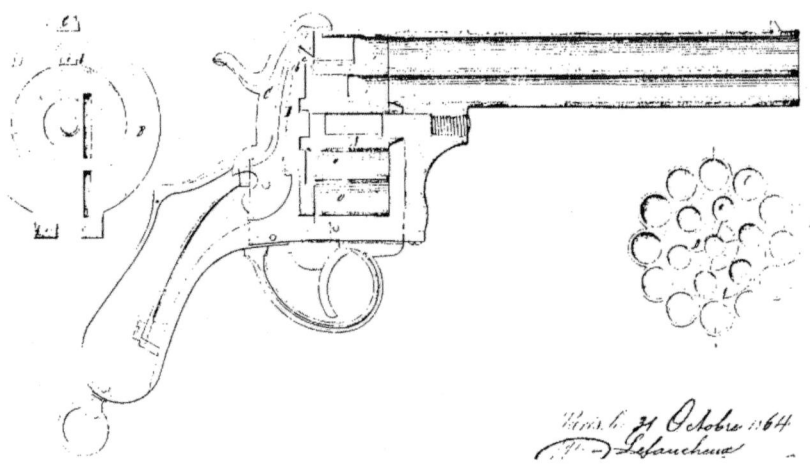

Technical drawing for patent 64960 of 17th December, 1864 showing the layout of the multi bore cylinder and the over and under barrels.  *Author's collection*

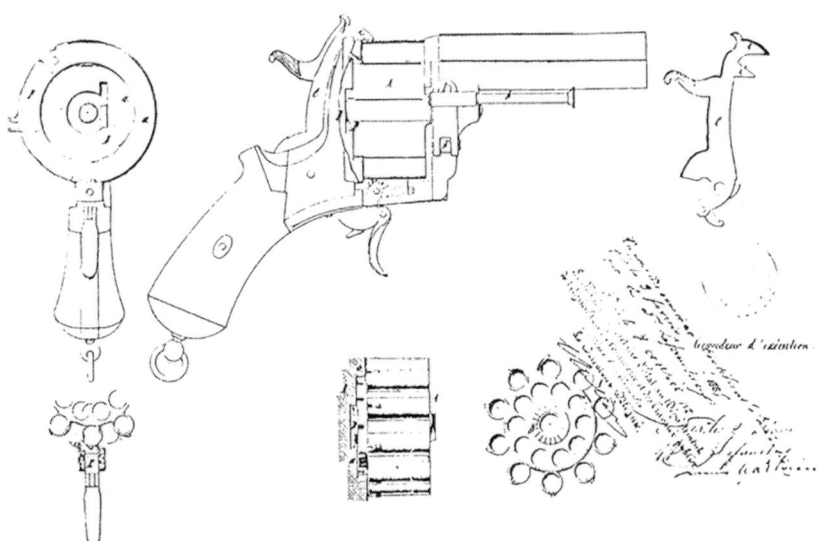

April 1865 first addition to patent 64960, showing the weight-saving fluted cylinder and the new hammer design are clearly shown.  *Author's collection*

weighing only a few ounces more than a standard 7mm 6 shot. These were very popular with around 2,000 being sold.

**1867** Lefaucheux attended the Paris Universal Exhibition where he displayed a new twenty shot revolving rifle with over and under barrels following the principle shown in his first addition to patent 64960, a sword revolver and a steel skeleton stock to turn a revolver to a carbine.

**1869** There were riots on the streets of Paris on the 8th and 9th of June following an election when the 'wrong' party won. Fearing that the 'White Overalls', as the ruffians were known, would try to arm themselves the new Cabinet ordered Lefaucheux, Devismes, Blanchard and all other principal gunsmiths to pull down their steel shutters and empty their shops of weapons. This proved to be good advice as on the evening of the 8th the mob spilled out of the Boulevard Montmartre and into the side streets. A gang of ne'er do wells attempted to break into Lefaucheux's shop on Rue Vivienne and only the arrival of the police foiled their attempt, several arrests were made.

Detail from a Lefaucheux handout from the 1867 Paris Exhibition showing the skeleton revolver stock which turns a long barrelled revolver into a carbine, and revolving rifles in 12mm and 15mm calibre. *Author's collection*

# PISTOLETS ET CARABINES REVOLVERS

### A SIX COUPS, SYSTEME

## E. LEFAUCHEUX

BREVETÉ S. G. D. G. (I)

Acceptés comme Armes réglementaires dans la Marine française, espagnole, sarde, etc. etc.

FABRIQUE ET VENTE EN GROS, RUE LAFAYETTE, 104, A PARIS.

Top half of the 1867 handout assuring customers that they "Make and sell wholesale".

*Author's collection*

**1870** The pinfire system reached its zenith as a military arm. Now government were calling for the new centre fire ammunition in their regulation side arms. The French Navy needed such a weapon to replace their old Model 1858's and Eugene obliged with a new design for a double action, closed frame, six shot cylinder chambered for the 11mm centre fire cartridge. Covered by patent 82358 4,000 were ordered on 10th February, but due to the intervention of the Franco-Prussian War and the Paris Commune the order was not acted upon until 1872. This was the last revolver that Eugene put into production deciding, rather to put all his efforts into long arms. He did continue to place patents for all arms though, ten patents and one certificate of addition between 1870 and 1878.

**1873** The Liege factory was sold to a Mr Dacier, but the Trade Mark was retained.

**1874** Lefaucheux bought an old factory called Tremerolles near his Chateau in Bruyers le Chatel, eighteen miles south west of Paris and an old mill on the nearby River Renarde which he refitted as gun barrel factory.

**1875** In June he closed his shop at 194 Rue Lafayette and moved in with another gunsmith, Jules Gevelot, at 32 Rue Notre Dame des Victoires.

**1878** The mill factory produced 4,400 37mm firing tubes for the French Navy's Hotchkiss guns. Another 1,000 were made in 1880.

**1881** On 15th November Eugene sold the Paris factory on Rue Vivienne, the goods and chattels stored in Notre Dame des Victoires and all his remaining patent rights to Chevalier and Dru. The deed of sale excluded him from any further involvement in the arms trade. He retired to his Chateau and concentrated on improving his farm.

**1891** Lefaucheux bought a villa in near Cannes on 25th March, whether for holiday or full time residency is not clear.

**1892** Eugene suffered a massive stroke on 2nd March and died aged 59 at his villa. He was interred next to his father in the family vault in Montmartre Cemetery. So ended the 75 year dominance of the Lefaucheux name in world gun making and innovation.

Casimir Lefaucheux 26th January, 1802, 9th August, 1859 inventor of the pinfire system. *Bibliotheque Nationale de France*

# Chapter Two

## PROOF MARKS FOUND ON PINFIRE REVOLVERS

### BELGIUM

Proof marks were first introduced into Belgium in 1672 under a voluntary system which worked very well for the Liege gunsmiths who had a thriving export trade, and would send out weapons to customers to finish and proof in their own countries. This all came to an end during the Napoleonic Wars when Belgium was annexed as a province of France. In 1810 Napoleon decreed that all weapons must carry proof marks on the St Etienne system. When, in 1830, Belgium again became an independent kingdom it was hoped they could return to the old voluntary system but the Napoleonic rules were only altered and continued as a mandatory law which came into force in 1836. Inspector marks were introduced in 1852-53 to ensure that weapons were not tampered with after the proof inspection.

   Definitive Liege proof mark
20th February, 1811 — 11th July, 1893

   Definitive Liege proof mark
11th July, 1893 — Present

   Inspector's mark
21st December, 1852 — 30th December, 1853

   Inspector's mark
30th December, 1856 — 26th January, 1877

   Inspector's mark
26th January, 1877 — Present

   Rifled barrel mark
1893 — Present

## FRANCE

French gunsmiths centred in St Etienne had been following a voluntary code of proof marking since the 1500s, with an official system being adopted in 1700. The law was not enforced until 1810 when Napoleon decreed it would become mandatory. In August 1855 the President of France was forced to abolish compulsory proofing by the courts. The gunsmiths, however, continued to use the facility as they were worried that customers would think that unmarked guns were of an inferior manufacture. A new proof house was opened in Paris in 1897.

St Etienne proof mark
19th February, 1824—30th April, 1856

St Etienne proof mark
30th April, 1856—22nd April, 1868

St Etienne proof mark
1869—1886

St Etienne proof mark
1886—30th July, 1897

St Etienne proof mark
30th July, 1897—1923

## GERMANY

Germany had a voluntary proof system from 1600 until the proof law of 1891 was passed by the Reichstag. This new law came into effect in April 1893 and required all firearms sold within Germany to be proof tested and marked on the barrel and frame. Exempted were imported weapons proof marked by a recognised authority.

   Proof mark applied to weapons in stock prior to new proof law in 1893

   Optional definitive proof mark 1893

   Definitive proof mark used on revolvers 1893—1939

## GREAT BRITAIN

In 1637 the Worshipful Company of Gunmakers of the City of London was formed and a proof house opened. In 1670 the view mark was added to show that the weapon had been inspected after the proof test. The workload at the London proof house eventually proved too much and so in 1813 the Company's powers were extended to all of England and Wales and a new proof house was opened in Birmingham to spread the load. In 1904 the laws were updated and new proof marks were introduced.

   London provisional proof mark 1637 to date.

   London view mark, 1670 to date.

   Birmingham provisional proof mark, 1813 to 1887

Birmingham view mark, 1813 to 1904

 Birmingham provisional proof mark, 1887 to 1904.

 Birmingham provisional proof mark, 1904 to 1925.

## SPAIN

A level of proof testing was carried out in Spain from the 16th century at the Royal Arsenal at Placensia, but this was not formalised until 1844 when a proof house was opened in Eibar. Proof testing was voluntary from 1844 until 1923 when it became a legal requirement, therefore very few pinfires were ever marked.

# Chapter Three

# EUROPEAN PINFIRE MANUFACTURERS AND PATENT HOLDERS

## BELGIUM

**ANCION**, Jaques et Cie. Gunsmith. Placed two patents for improvements to Lefaucheux revolvers, one in 1887 and one in 1888. Trade mark is crown over JAC.

**ANCION–MARX**, Leopold. Gunsmith, 26-28 Rue du Grandgagnage, Liege, then numbers 26-28 from 1933. Registered from 1897 to 1947. Registered seven patents between 1897 and 1933.

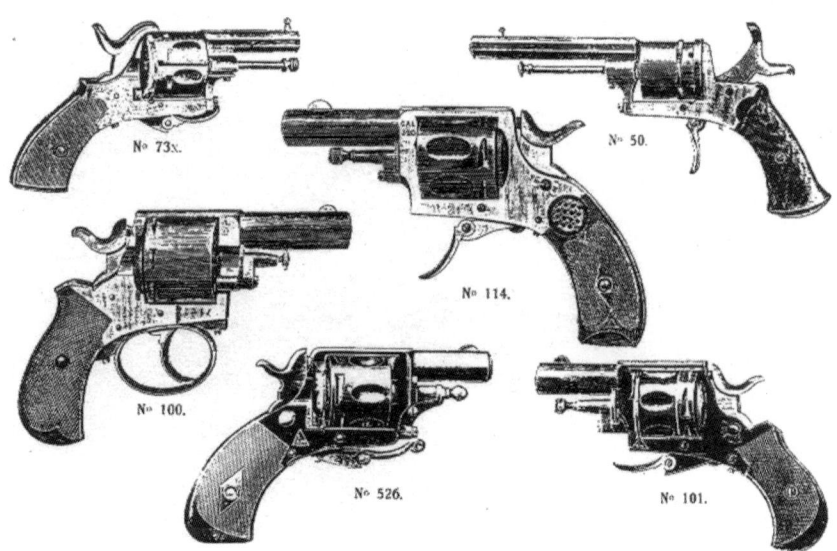

A page from the 1909 Ancion-Marx catalogue, item 50 shows that pinfire pistols were still being made, although the rim fire Bulldog type had taken over in popularity.   *littlegun.be*

**ARENDT**, Maurice. 8 Rue Trappe, Liege. Gunsmith, registered with the Belgian proof house between 1857-89. Granted patents for improvements to Lefaucheux revolvers and for a method of crimping pinfire cartridges. Trade Mark is crown over MA.

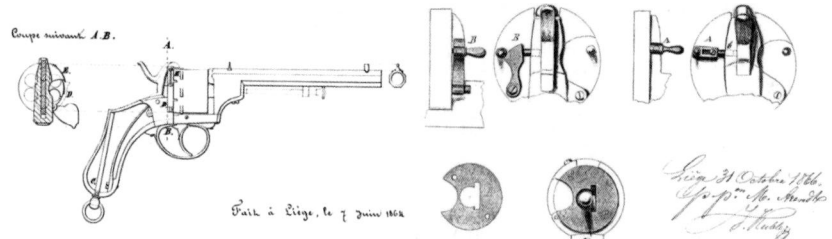

Arendt patent dated 31st June, 1864 for a new style of loading gate. Another patent, dated 31st October, 1866 is for two safety devices to prevent an accidental discharge. The first is the addition of a rim to the recoil plate which covers the exposed pins protruding from the cylinder, in the second the hammer is prevented from dropping on the cartridge pin by the use of a sliding bar which holds it just above the pin.

*Author's collection*

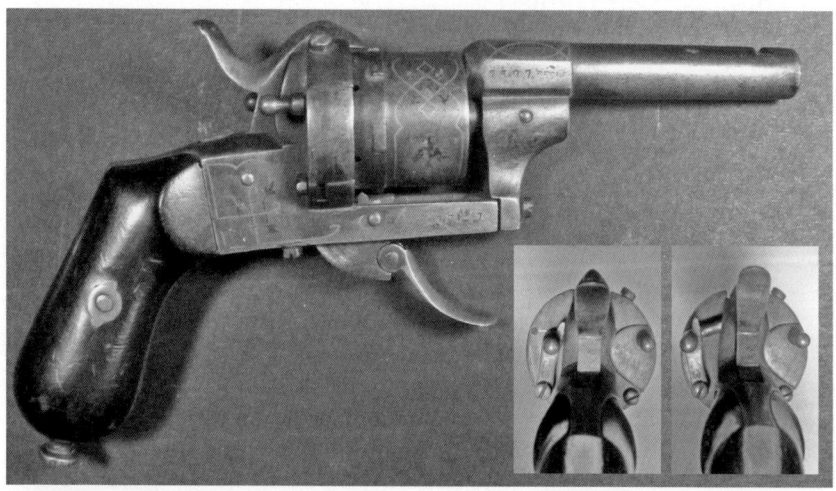

c.1866 7mm pinfire by Maurice Arendt showing all of the improvements in the patents above. A pin shield and drop down loading gate and sliding safety catch. Inset, rear view of the weapon showing the safety catch in the on and off positions. *Author's collection*

**BAYET FRERES,** Gunsmiths, registered 1865-73. Between 1865 and 1873 they deposited six patents for the improvement of revolvers. Trade mark B F divided by crossed swords.

**BERNARD**, Victor. Liege. Gunsmith, registered at the proof house 1873-81. Placed twelve patents between 1872 and 1888 including pinfire cartridge improvements.

**BERNIMOLIN**, Victor. Gunsmith, 2 Rue Des Chapelains, Liege. Registered 1879-82.

**BERNIMOLIN FRERES**, Nicholas and Victor in association with H.J Cap. 2 Rue des Chapelains, Liege. Registered at the proof house 1882-85.

**BERTRAND**, Antoine et fils. Gunsmith, 25 Rue Fabry, Liege. Was awarded patents in 1886 and 1887. Introduced the naming of weapons like The Hunter, The Bold and the Defender. Trade mark is crowned B&F.

**CHARLES**, J. Liege. Two patents in 1858 for pinfire cartridges.

**CHAINEUX, J.** Based in Wandre, a district of Western Liege. Gunsmith, c.1858-64.

**CHAINEUX**, Joseph Lambert. Gunsmith, 26 Place des Carmes, Liege. Registered 1864-84, placed patents in 1863 and 1864.

**CLEMENT**, Charles. Gunsmith, 37 Rue Cheri, Liege. On the proof house register from 1883-1912. Deposited 35 patents and registered marks between 1885-1912, including 'The Guardian American Model of 1878' registered on 27th December, 1880, and the snappily titled 'The New English Pattern Pinfire Pistol 7mm 20th September, 1876'. Clement seems to have marked very few, if any of his work with an identifying stamp and so we must rely on the various names he gave his weapons for examples. Charles also worked in collaboration with Alexandre Fagnus. See **FAGNUS et CLEMENT**.

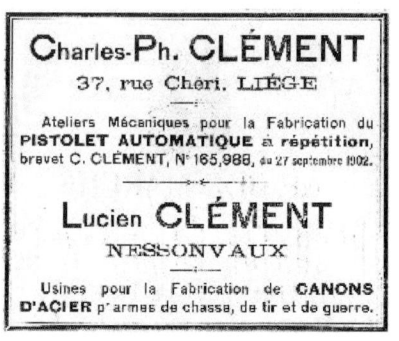

An early 20th century advertisement for Clement's gun shop on Rue Cheri.
*Christain Feron*

**CLEMENT et DOLMAN**, 5 Rue Forgeur, Liege. Following the death of her husband in 1913 the widow Clement nee Dolman renamed the company. This lasted only until 1914 so perhaps it was only to cover the period needed to dispose of old stock and patent rights.

7mm snub nose revolver by Clement. Identical to the later 'New English Pattern Pin Fire' except that in the latter the extractor rod is housed in the butt.  *Author's collection*

Unmarked other than the standard proof marks these two 7mm closed frame revolvers are definitely, from the Clement stable. Both have pre-1893 proof marks and post-1877 inspector's marks. The New English Pattern Pin Fire Pistol 7mm 20th September, 1878 is a cheap 'Saturday Night Special' style weapon.  *Author's collection*

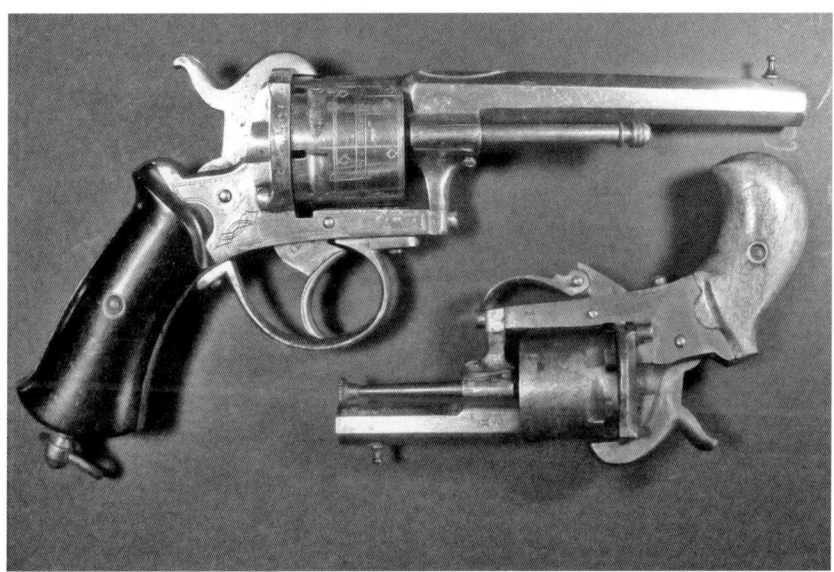

The 'Guardian American model of 1878' in the scarce 9mm calibre. It has the title on a ribbon round the cylinder. Post-1877 inspector's mark and pre-1893 Liege proof mark consistent with its registration date of December 1880. Bottom. A 7mm calibre Guardian, these were also available in 12mm but they are rare. *Author's collection*

**CLOSSET,** Joseph. Gunsmith, 231 Ruelle en Glain, Liege. Registered trader 1871-88. Trade mark J.C.

**CLOSSET,** Louis. Gunsmith, 72 Rue Hors-Chateau, Liege. Registered trader 1896-1921. Trade mark circled L.

**COLARD,** Louis. Gunsmith, 68 Rue Louvrex, Liege. In 1865 registered a patent for a revolver. Trade mark crowned L.C.

**COLLETTE,** Victor. Gunsmith, 60 Quay St Leonard, Liege. Active 1836-1909. Trade mark with star, or plain VC.

**COMBLAIN,** Cheratte, liege.

**COQUIHAT et DIGNEFFE,** Gunsmiths, registered 1855-58. Trade mark C&D in oval.

**CROUTE**, Theodore. Gunsmith, Mortier, Liege. Registered 1864-80. Trade mark star over TC.

**COURARD**, Hubert Joseph. Gunsmith, Argentau, Liege.

**DARRIEN**, V. Liege. Patent holder 1876.

**DECORTIS**, Francois Joseph. Sarolay-Argenteau, Liege. Placed four pistol patents between 1863-66.

**DEGUELDRE**, Constant Joseph. Gunsmith, St Remy. On 26th July, 1866 Degueldre applied for a patent for a device for the quick ejection of spent pinfire rounds. Pulling down a lever causes the barrel and cylinder to move forward on the cylinder pin. A ring fixed to the recoil plate and forward of the pin cartridge holds back the cartridge case as the cylinder moves forward, causing the empties to drop out all at once.

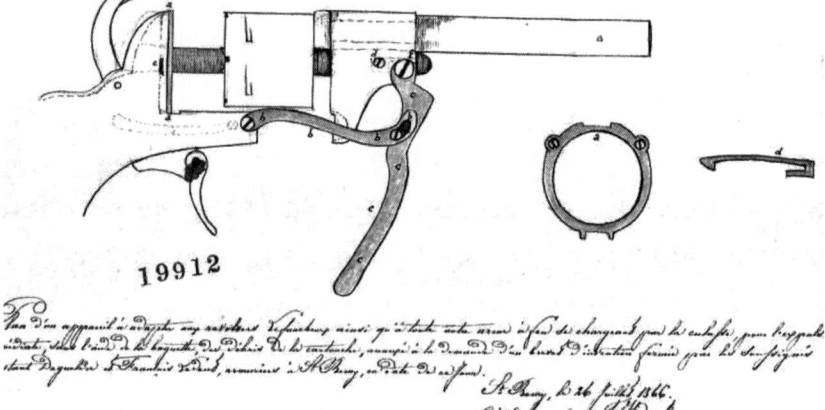

Degueldre's patent 19912 of 1866 for the swift ejection of spent pinfire cases. The diagram shows how the dropping of the lever pulls the cylinder and barrel along the cylinder pin, the ring catches the protruding pins and removes all the spent cartridges in one go.

*Author's collection*

# European Pinfire Manufacturers and Patent Holders

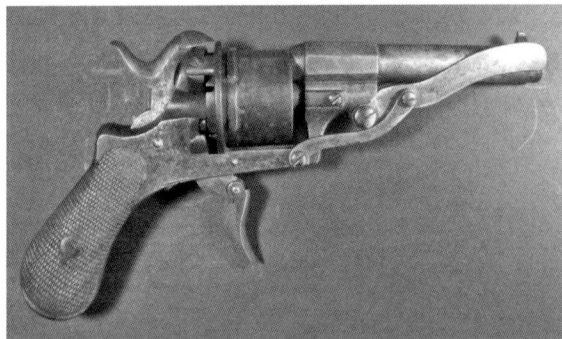

A double action 7mm Degueldre patent revolver made by E Lefaucheux in Liege in 1868 and proofed in Birmingham. As the lever is lowered the cylinder and barrel assembly move forward so that the fixed ring ejects the spent cases.
*Author's collection*

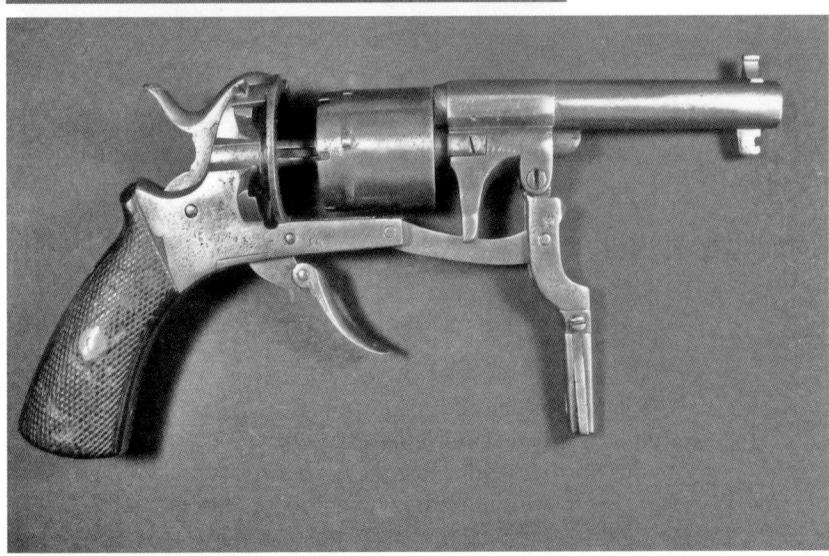

A later, but still pre-1877 Liege made 7mm Degueldre patent. This version has the lever fixed under the barrel by a sprung clip. *Author's collection*

**DEPREZ**, Jean Mathieu. Gunsmith, 380 Rue del la Chapelle, Herstal, Liege. Holder of seven revolver patents issued between 1857-67. Best known for his pepper-pot revolvers.

**DEUSTER**, Guillaume. Liege. Registered with the proof house between 1860-78, then 1879-85.

**DOLNE**, Louis. Gunsmith, 8 Rue Stephany, Liege. Registered 1873-81. Best known for his 'Apache' combination pinfire pepper-pot pistol, knife and kuckleduster. Dolne claimed to be the inventor of this style of weapon, although the Delhaxe patent below proves otherwise. The Apache is named after the brutal Paris gang infamous for their rough way of fighting.

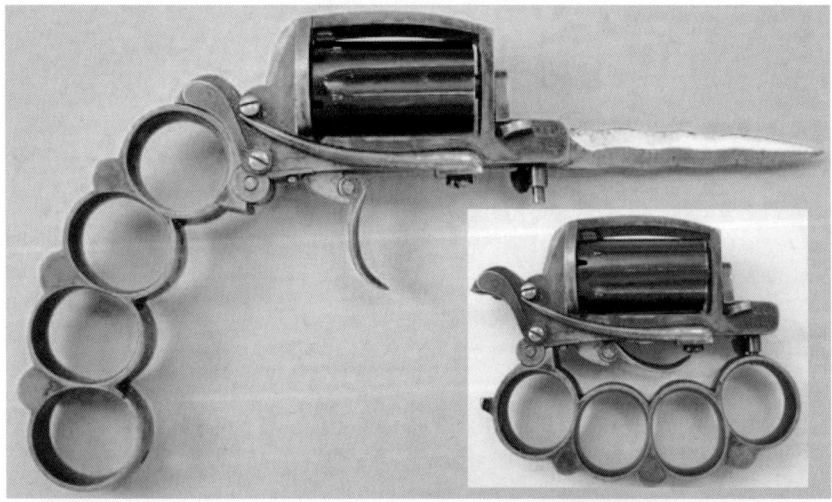

A Dolne 'Apache' shown in the open gun-knife combination and closed as a knuckleduster. *Horst Held*

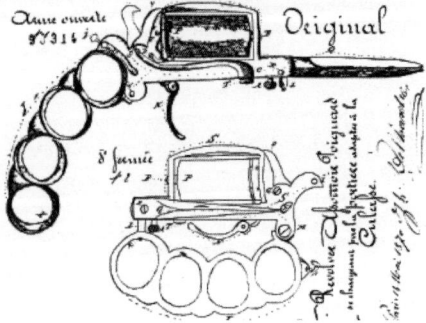

Delhaxhe's French patent 90314, 4th May, 1870 shows the origin of Dolne's combination weapon.
*Author's collection*

## European Pinfire Manufacturers and Patent Holders

**DRESSE, ANCION, LALOUX et CIE,** Gunsmiths, 47 Sur la Fontaine. This temporary amalgamation was put together to fulfil larger orders that the individual gunsmiths could not manage on their own. Registered at the Liege proof house between 1862-66. Known to have produced pinfire revolvers. Trade mark is D.A.L & C$^{ie}$.

D.A.L & C$^{IE}$

**DRESSE-LALOUX,** Gunsmiths, 3 Rue des Urbanistes, Liege. A permanent pairing of two of the gunsmiths mentioned above, they were registered at the Liege proof house from 1867-1908. Their trade mark is a bee with the letters D L on the wings.

Dresse-Laloux advertisement dating to c.1900.   *littlegun.be*

**DRISSEN,** Ferdinand. Gunsmith, 10 Rue Forgeur, Liege. Registered at the proof house 1874-1900. Trade mark FD in a lozenge.

**FAGNUS,** Alexandre, et Cie. Gunsmith, 9 Rue des Celestines, Liege. Registered at the proof house 1870-79. Owner of seven revolver patents.

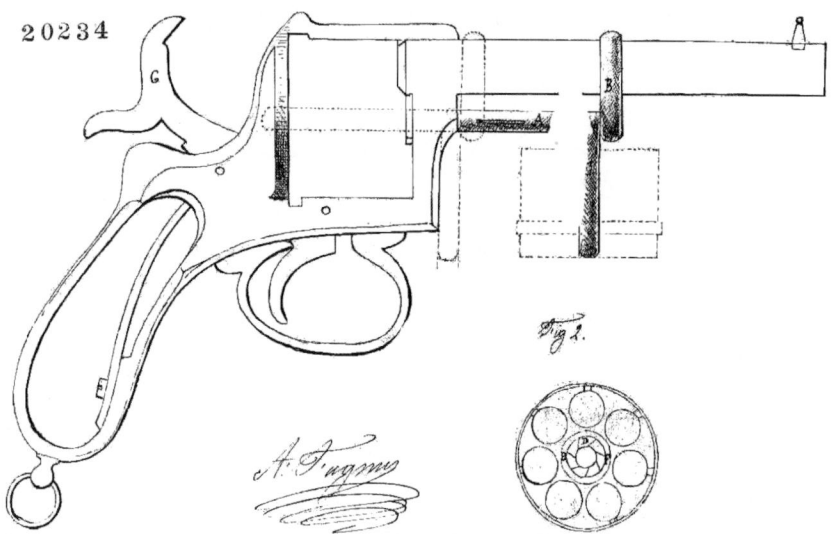

Fagnus patent 20234. Sliding the bar and ring assembly along the barrel withdraws the cylinder pin and allows the cylinder to drop out of the frame allowing for quick loading and unloading.  
*Author's collection*

12mm Fagnus patent number 20234 double action revolver, proof marked in Birmingham.  
*Author's collection*

EUROPEAN PINFIRE MANUFACTURERS AND PATENT HOLDERS

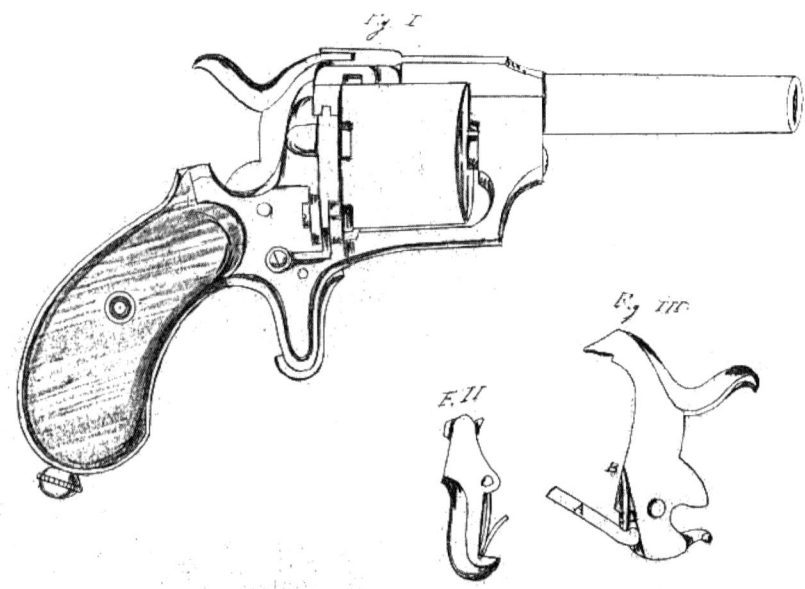

Fagnus patent number 30174 for a solid framed, spur triggered revolver in the American style, perhaps to raise business in the USA. I have only seen four of these for sale in the last 30 years and two of them were in American auctions. *Author's collection*

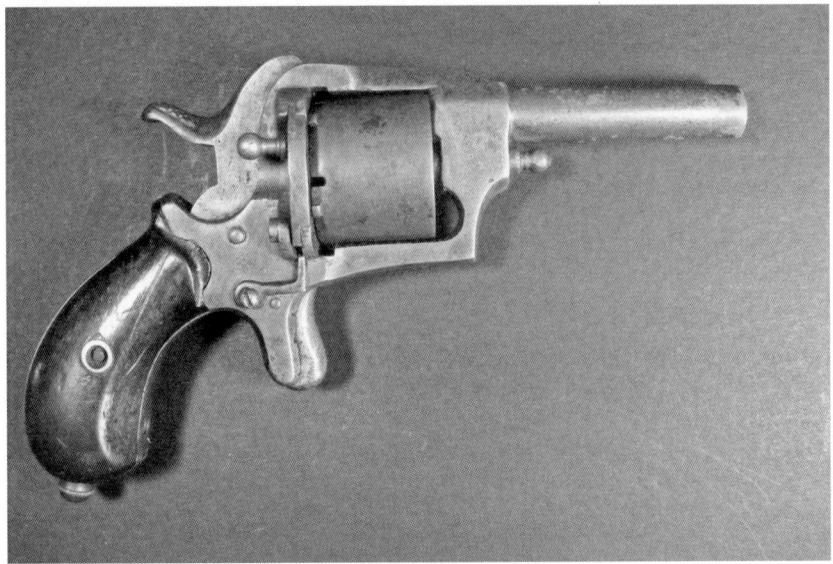

A 7mm spur triggered revolver. Apart from the Liege proofing it is completely unmarked, but it is without doubt a Fagnus product. *Author's collection*

**FAGNUS et CLEMENT,** Gunsmiths, 37 Rue Cheri, Liege. Alexandre Fagnus and Charles Clement entered into a partnership in 1879 sharing patents and production space in Clements premises on Rue Cheri. The partnership ended in 1883. Trade mark is F & C in crossed pistols.

**FLIEGENSCHMIDT,** Max. Gunsmith, 30 Rue Courtois, Liege and later 30 Rue du Vivier, Liege. Trade mark is crowned MF.

**FRAIKIN,** Jacques Joseph. Gunsmith Trembleur, Liege. c.1862-65. Awarded four patents, two of which were for pinfire revolvers.

**FRAIPONT,** Emile. Gunsmith, 45 Rue Monulphe Liege. Registered 1903-27. Trade mark fancy EF.

**FRANCOTTE,** Auguste. Gunsmith, 61 Rue Mont St Martin. Liege. Registered at the proof house 1848-68. Trade mark crowned AF.

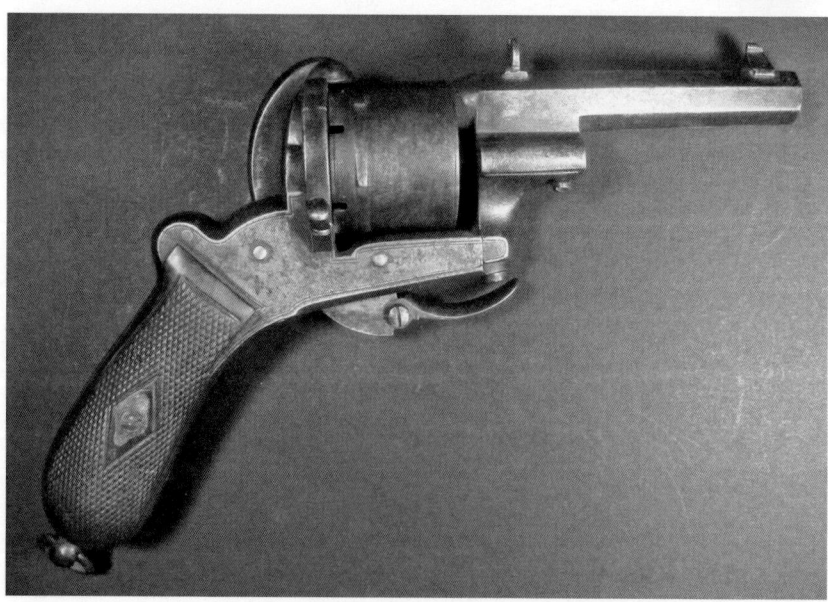

6 shot 9mm Lefaucheux Brevette self-cocking pinfire revolver by Francotte.
*Author's collection*

**GROSFILS**, Dieudonne. Gunsmith, Wandre. Three patents placed between 1864-65. Mark stamped on butt frame under grip.

**GULPEN**, Hubert. Gunsmith, 27 Quai de L'Abatoir, Liege. Registered at the proof house between 1910-14. Trade mark is circled G

A pair of standard late 7mm pinfire revolvers produced by Gulpen just before the First World War. Virtually identical except the lower one has a more rounded barrel lug and the barrel itself is 5mm longer. *Author's collection*

**HENKET**, N J. Gunsmith, Vivegnus, then Liege. Placed four patents between 1859-84. Trade mark NJH. **NJH**

**HERMAN**, Jean Jacques. Cheratte, Liege. Patents for pistols and revolvers in 1837, 1839, 1854, 1857 and 1860.

**HAAKEN**, Charles, et Cie, Gunsmith, Rue Jonfosse, Liege. Active during the period 1859-94. Renamed **HAAKEN et Fils** between 1894-98. Trade mark is initials CH circled.

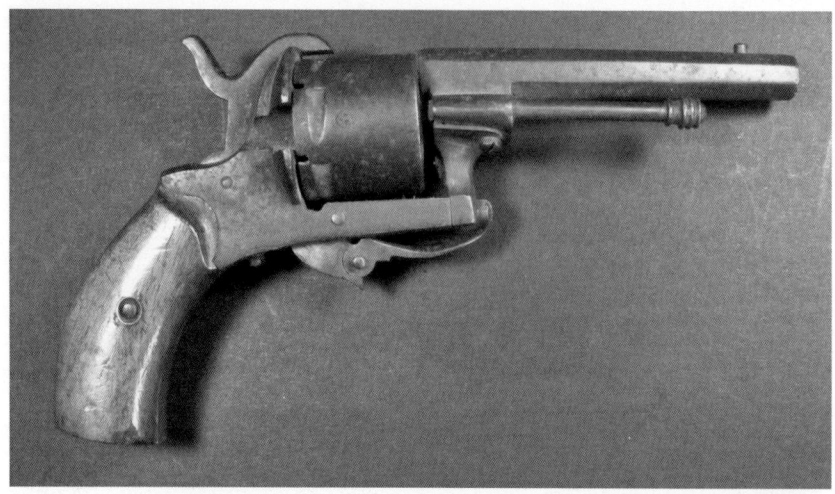

Cheap and cheerful 7mm 6 shot revolver by Haaken et Cie. The loading gate is not missing, it was never there. A simple gap in the recoil plate is provided for loading and unloading.
*Author's collection*

**HERMAN-LEDOUX,** Joseph. Gunsmith, 70 Rue de L'eglise Chenee, a district of Liege in 1866, then moved to Herstal in 1867. Deposited four patents in 1866 for pepper-pot revolvers. Trade mark is JHL.  **JHL**

**HONHON,** Alexandre. Gunsmith of Liege. Registered 1868-77. Three patents to improve Lefaucheux rifles.

**JANSEN,** Adolphe. Gunsmith, 27 Rue Madeleine, Brussels. Registered at the Liege proof house 1859-84. Placed nine patents including one for improvements to the Lefaucheux rifle. Trade mark is crowned AJ.

**JACQUEMART,** Joseph et fils. Gunsmiths, Liege. Known to have produced double barrel pinfire boot pistols. Placed one patent in Belgium. (JJ)

**JONGEN FRERES,** Gunsmiths, Liege.  Registered at the Liege proof house between 1856-73 when it was sold to Jamart-Smits. Deposited four revolver patents.

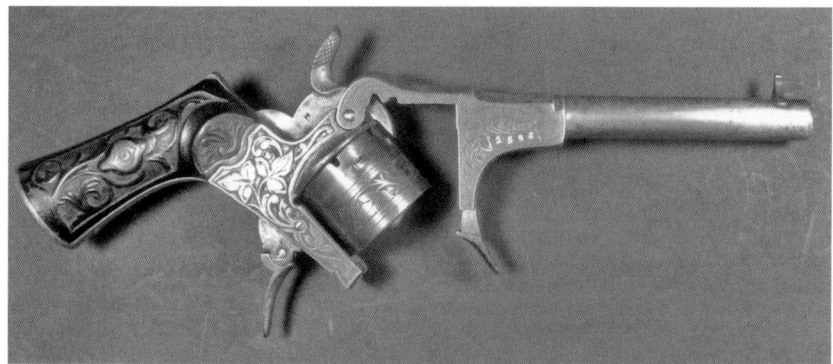

Jongen Freres patent tip revolver. This 7mm esample is shown in open and closed positions.
*Author's collection*

**JULIEN,** Joseph. Gunsmith, 3 Rue Nagelmackers, Liege. Deposited five patents between 1865 and 1876. Trade mark crowned JJ.

♛
**JJ**

**KAUFMANN,** Joules et Cie. Gunsmith, 65 Rue Jonfosse (1890), 127 Rue Jonfosse (1894), finally 26 Rue Augustines in 1899. From 1918 known as Lepage Arms Manufacture. Registered eighteen gun brand names between 1890 and 1918.

**JK&C°**

**KINAPEN,** Alfred and Francois. Gunsmiths in Liege. Registered at the proof house as F Kinapen et Cie 1862-63 and F Kinapen 1863-73. Deposited revolver patents in 1855 and 1861.

**LAIRESSE**, Louis (father) and Louis (son). Gunsmiths, 76 Rue St Severin, Liege (1893-94) then 249 Rue St Giles, Liege (1903-04). Registered at the proof house 1885-86. Given a patent for improving Lefaucheux rifles 1893.

**LAISKIN**, L. Gunsmith, 226 Rue St Giles, Liege.

**LAPORT**, Guilliaume. Gunsmith, 35 Quai St Leonard, Liege. Registered at the proof house 1850-63, then as Laport et Cie 1863-1900.

**LARON**, Pierre Antone. Gunsmith, 20 Place du Grand Sablon, Brussels. Patent holder.

**LEDENT**, Francois Joseph. Gunsmith, St Remy, Liege. In collaboration with Degueldre placed five patents between 1864 and 1870 for the safe ejection of spent cases from Lefaucheux pistols. Trade mark crossed swords, crowned, over FL.

A fairly standard, lower end Ledent 7mm double action revolver with the only nod to quality being the cross-hatched grips. He could though produce quality weapons with fine engraving. *Author's collection*

**LEERS, N.** Trembleur, Liege. Given a patent for a breech-loading Lefaucheux type pistol in 1862.

**LEFAUCHEUX et Cie.**, Gunsmiths, 13 Quay de Fragnee. 1862-69. Early weapons produced at the Liege factory carried the Broken Gun trade mark as used on Paris-made guns, later the crowned LF was used, sometimes with the year of manufacture underneath.

**LEJUNE, J.** Gunsmith, 100 Rue Louvrex, Liege. Awarded patents in 1885 and 1906. Trade mark is crowned JL.

**LEONARD, L.** Gunsmith, 19 Rue des Gris, Herstal, Liege.

**LEPAGE et CHAUVOT**, Alphonse and Frederic. Gunsmiths, 25 Rue Fabry, Liege Registered 1867-81 Trade mark is L&C.

**LOVINFOSSE-HARDY et Fils.** Gunsmith, Rue Hayeneux, Herstal, Liege. Active between 1878 and 1890. Deposited two patents, one in 1885 for a new safety device for breech-loading weapons and a second for an extractor for single shot, breech-loading pistols in 1909 long after they ceased trading.

9mm double action revolver from Lovinfosse-Hardy. Marked with pre-1893 proof mark and post-1877 inspector's marks this dates from the mid 1880s. *Author's collection*

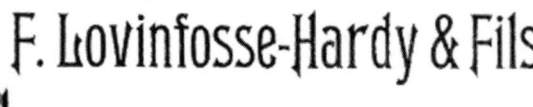

## F. Lovinfosse-Hardy & Fils

HERSTAL (Belgique)

Advertisement for Lovinfosse-Hardy & Sons "Arms maker for War, Luxury and Export.
*Cristian Feron*

**MAIRLOT**, Guilliaume. Liege. Awarded a patent in 1872 for a system for firing centre fire ammunition from Lefaucheux revolvers.

**MALCHAIR**, H. Gunsmith, Cheratte, Liege. Patent awarded in 1863 for a mechanism to close the breech and eject spent rounds from pinfire revolvers, and another in 1867 for needle fire weapons.

**MARIETTE**, Guillaume. Gunsmith, Cheratte, Liege. Registered at the proof house from 1832 to 1865, he was awarded 22 Belgian patents between 1840 and 1888. Trade mark is MG crowned.

**MARTIN**, Jacques Francois. First found in Dahlem, a district of Liege in 1866, then 31 Rue Gerardie, Liege From 1884-85. He received three patents, one of which was for improvements to Lefaucheux revolvers.

**MERLOT**, N. Ans et Glain. Awarded a patent in 1869 for improvements to Lefaucheux rifles.

# European Pinfire Manufacturers and Patent Holders

Detail from Mariette's patent dated 21st May, 1840 for a six shot under hammer revolving percussion pistol. The usual ring trigger which is characteristic of most of Mariette's work is not shown on this drawing. This style of revolver was copied by Casimir Lefaucheux for the introduction of his pinfire system.

*Author's collection*

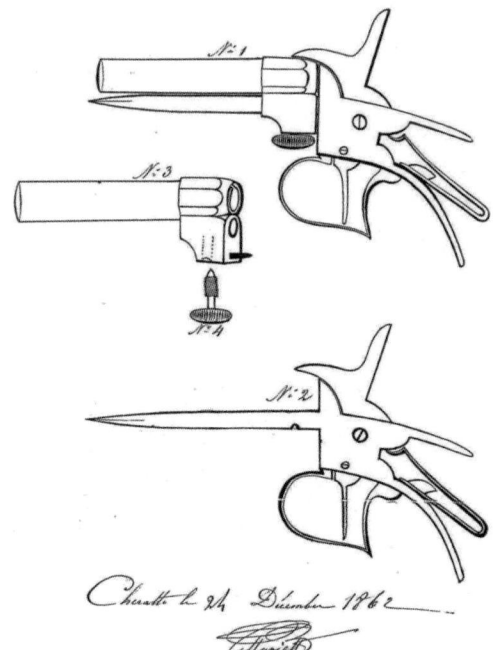

Two details from Mariette's Belgian patent 22nd December, 1862. The one above is for a single shot pistol which is loaded and unloaded by removing the barrel and using the pin shaped frame extension to push out the empty case.

*Author's collection*

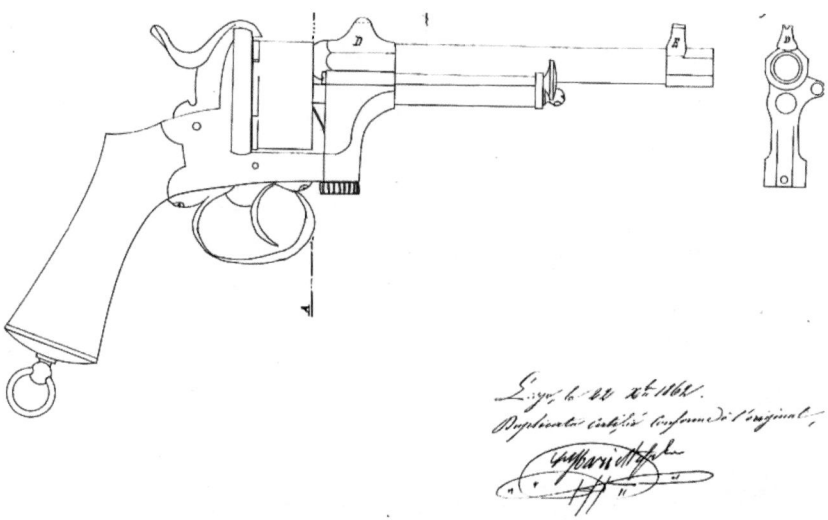

Mariette patent for a revolver which can be quickly loaded or emptied by removing the barrel and sliding off the cylinder.
*Author's collection*

A finely engraved, 7mm Mariette December 1862 Brevet double action six shot revolver in the ready to fire postion. Unscrewing the knurled nut beneath the barrel assembly releases it and allows you to slide it off, this then allows the removal of the cylinder for loading and unloading. The cylinder pin is used to push out spent cartridges. Once the cylinder has been reloaded the process is reversed to ready the gun for firing.
*Author's collection*

**MEROLLA FRERE**, Giovanni and Francesca. Gunsmiths, in Liege. Active c.1860-70, they also had a factory in Naples, Italy.

**MEYERS**, Guillaume Joseph. Gunsmiths of Cheratte, Liege. Awarded seven patents between 1863 and 1877 for revolver innovations.

Meyers patent for a quick loading device. Top. The weapon, a 7mm double action solid frame revolver ready to fire. Below. Pulling the slide attached to the cylinder pin along the barrel until a detent is reached allows the cylinder to be released from the right-hand side of the frame. Unloading is achieved using the ejector rod as usual. *Author's collection*

**MICHAUX**, P N of Herstal. Liege. Issued three patents between 1858-65, the last one for improvements to the breech gas seal on Lefaucheau rifles.

**MOTTET**, H. 3 Rue Degres des Tisserands. Awarded a patent in 1873 for improvements to Lefaucheux arms.

**MULLER**, Louis. Gunsmith, 14 Place de Bronck. Active c.1889-94. Owner of twenty trade marks issued between 1889 and 1894. Trade mark is a five point star.

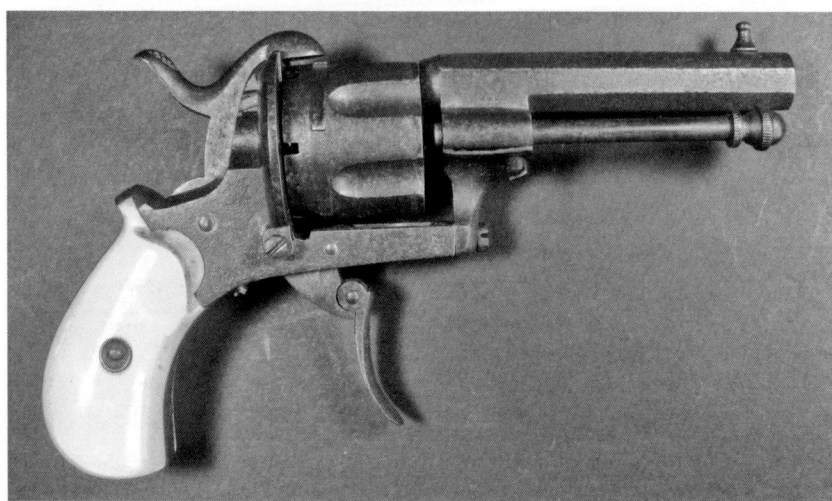

Muller 7mm revolver. Typical cheap, mass produced pinfire of the late 19th century but taken out of the ordinary by the use of the old fashioned bird's head grip with faux ivory, resin grips and the half fluting of the cylinder. *Author's collection*

A Muller 7mm revolver with finely moulded gutta-percha grips. It is in its oversize travelling pipe case which has space for a dozen spare cartridges. *Author's collection*

**NAGANT**, L. Gunsmith, 89 Fauborg St Gilles, Liege. Awarded four patents, the first for modifications to Lefaucheux rifles, the second a design for metal pinfire cartridges. The third was for further improvements to metal pinfire cartridges and the final one was for a system to make a revolver into a carbine.

**NEUMANN FRERES.** Gunsmiths, 7 Rue St I Remy, Liege. Active *c.*1893-1948. Awarded nine patents between 1890-1905 including one for refinements to the Lefaucheau rifles.

**OLIVIER**, Gilles. Gunsmith, Herstal, Liege. Registered 1868-90. Awarded 40 patents including one to convert Lefaucheux revolvers to centre fire.

**OLIVIER**, Gilles Noel. Gunsmith, Herstal, Liege. Registered at the proof house 1854-61. Holder of five Belgian patents including one for modification to Lefaufeux rifles.

**OLIVIER**, Pierre, Gunsmith, Herstal Liege. Holder of three patents, all for improvements to Lefaucheux weapons.

**ORTMANN**, Hermann. Gunsmith, 13 Rue D'amay, Liege. Registered 1866-69. Manufactured guns under names such as 'Thunderer New American Model', 'The LightningWorlds Fair 1878' and 'The Defender American Model of 1878', trade marks for these three being issued in January 1887.

**OURY**, J J. Gunsmith, located in St Remy, Liege. Awarded four patents between 1859-64. The 1859 patent, number 8300, is for a quick release assembly very similar to the Mariette patent.

**PEREE**, Arthur Louis Joseph. Gunsmith, 33 Rue Basse Sauveniere, Liege. Active between 1884-91.

**PIRE**, Laurent. Gunsmith, from Liege, active 1856-58.

**PIRLOT FRERES,** Gunsmiths, 95 Rue St Gilles and later 39 Avenue D'avroy, Liege. Registered at the Liege proof house between 1830-79. Holders of three Belgian patents they tended to concentrate on high quality copies of other peoples' designs manufactured under licence. In 1879 they merged with Fresart, becoming Pirlot et Fresat. This pairing lasted until 1890 when they were absorbed by Societe Anonym Fabrication d'Armes a Feu.

A top quality 12mm double action Pirlot Freres revolver. The proof and inspection marks date it prior to 1877 and the general style places it firmly in the 1860s.  *Author's collection*

A 12mm revolver by Pirlot Freres dating to the mid 1860s. This is one of many pinfire revolvers exported to Argentina by Pirlot. Unfortunately the retailer's name is so worn it is unreadable but the Buenos Aires address is clear.  *Author's collection*

# European Pinfire Manufacturers and Patent Holders

Another finely engraved Pirlot Freres 12mm revolver sent to South America, this one dates to the early 1870s. The Argentinian Gauchos preferred to have large calibre weapons with as much embellishment as possible and this one ticks all the boxes. Fine foliate engraving covers the frame, cylinder, trigger guard and butt plate. This particular gun was imported by L Paris of 20 Calle Rivadavia, Buenos Aires and his name is beautifully engraved on the top strap. Paris was a retailer of top quality weapons and his cased weapons often appear on sale in the UK. *Author's collection*

An 1880s advertisement for Buenos Aires arms retailer Carlos Rasetti of 526 Rivadavia, the same street as L. Paris, showing a French double barrel boot gun. *Author's collection*

**PLIERS**, Adrien. Gunsmith, from Liege. Entered six patents between 1855-66, known to have made pinfire revolvers which are marked A. PLIERS BREVETE.

**POLAIN**, Prosper. Gunsmith, 243 Rue St Leonard, Liege. Author of eleven Belgian patents between 1856-73.

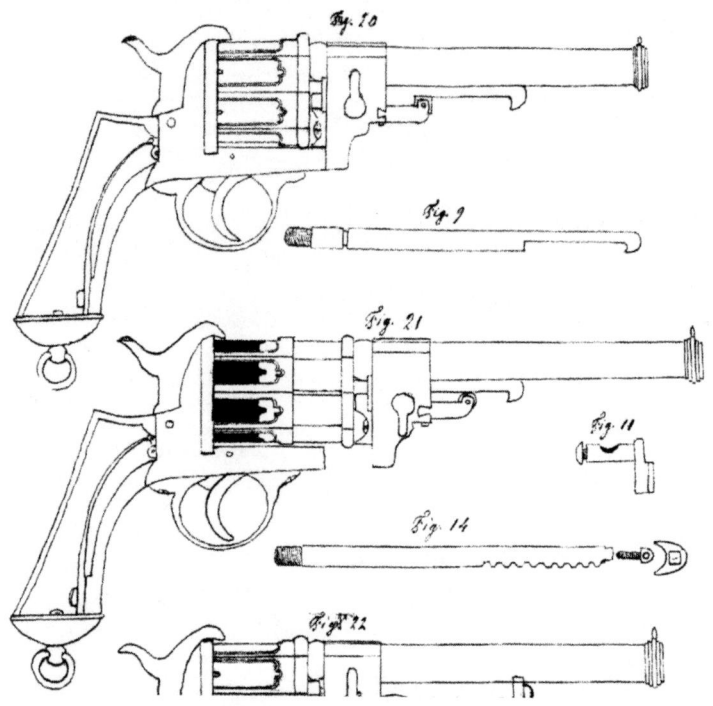

Detail from an early Polain patent showing a very ornate revolver with a removable cylinder pin to enable fast loading and unloading. With the cylinder pin removed the cylinder drops out and can quickly be loaded or unloaded. The cylinder pin is used to eject spent cases.
*Author's collection*

**POOT**, Gustave Gunsmith, of Liege. Active 1856-85. Produced one patent in 1857 applicable to Lefaucheux weapons.

**POULAIN**, L Gunsmith, from Ans, Belgium. Awarded a patent in 1875 for an improvement to Lefaucheux weapons.

**REEL**, J J. Gunsmith, 2 Rue du Tir, Liege. Active in 1913, known to have produced pinfire revolvers.

**RENKIN et Fils**. Gunsmith, 86 Boulevard d'Avroy, Liege. Active c.1896-1948. Known producer of pinfire revolvers. Trade mark is a lion rampant holding a shield with crowned R on it.

**RENNOTTE**, Dieudonne Joseph. Gunsmith and lock maker, Housse-lez-Liege. Registered at the Liege proof house 1846-67, and deposited five patents.

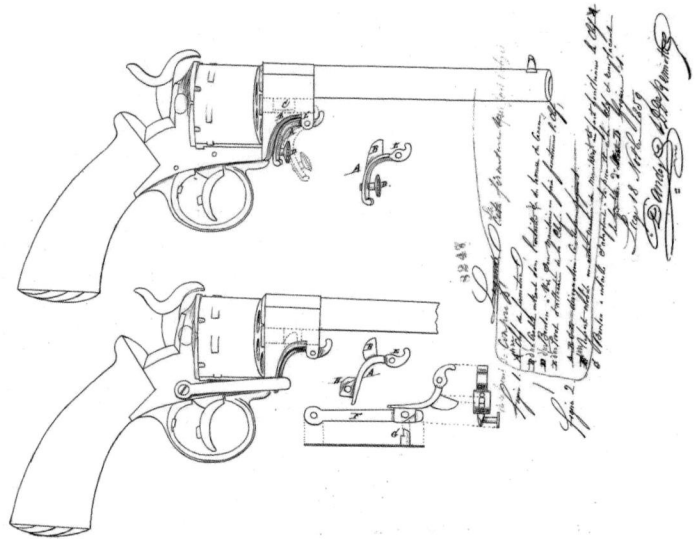

Rennotte patent number 8247, date of application 18th November, 1859. Another quick loading/unloading patent, doomed to never go into serious production. The top drawing shows a key held in position with a knurled nut which when unscrewed allows first the barrel then the cylinder to be removed. The lower drawing shows that the cylinder drops out and can quickly be loaded or unloaded; the cylinder pin is used to eject spent cases.

*Author's collection*

**RIGA**, J J. Gunsmith, 282 Rue Vivegnis, Liege. Deposited a patent application in 1860 for improvements to Lefaucheux rifles.

**RISSACK**, N J. Gunsmith, Hoignee, Liege. Awarded a patent in 1858 for improvements to the gas tight seal for Lefaucheux rifles and othe makes.

**ROCOUR**, B. Gunsmith, 13 Rue Sainte Foi, Liege. Active 1872-75. Awarded a patent for a type of barrel for Lefaucheux revolvers.

**RONGE, Jean Baptiste et FILS.** Gunsmiths, 4 Place St Jean, Liege then Place Xavier Neujean. On the Liege proof house register from 1832-1939. Awarded five patents between 1894-1928. Trademark is the initials JBR crowned or the same between crossed swords.

A fairly standard 7mm revolver by Jean Baptiste Ronge et Fils from the time when the sons had taken over the business. Dating the weapon becomes difficult when the proof house marks are an impossible combination. The proof mark of a crowned circle with ELG star is for the period 1893-present. The inspector's mark of a crowned R was in use during the period 1853-77 giving us a sixteen year gap which should have been filled by the use of the new star R mark. The answer is quite simple though; the crowned R tells us that this gun has a rifled barrel. The fancy R mark was introduced in 1894 for use on all rifled handguns.

*Author's collection*

**ROUX, Jacques.** Gunsmith, from Liege. Registered at the Liege proof house between 1872-90, but was active in the gun trade before this as he began placing patents from 1840 through to 1873 with a total of ten. Two of the patents were for improvements to Lefaucheux long arms.

**SAUVEUR, Freres.** Gunsmith, 103 Rue Laixhaut, Herstal,Liege. Placed four patents between 1884-87.

**SAUVEUR, Henri.** Gunsmith, registered at the Liege proof house 1903-08. Became Sauveur et Fils in 1908.

SAUVEUR, Henri et Fils. Gunsmiths, 9 Rue Hors Chateau, Liege. Registered at the proof house from 1908-32. Also owned a factory and outlet at 2 Marche aux Oiseaux, Ghent. Trade mark was H.S in a lozenge.

H Sauveur advertisement for pocket revolvers. It is nice to still see a pinfire revolver amongst the Bulldog and Veolodogs.   *littlegun.be*

SAUVEUR, Jean. Gunsmith, 10 Rue de Bassenge. Active 1905 until 1910 when he joined his brother Henri. He left in 1923 and resumed working under his own name until 1933 when he rejoined the family business. His trade mark was his initials JS.

SAUVEUR, Henry Joseph. Gunsmith. (This Sauveur and the three below are a different branch of the family above). Awarded a patent to improve Lefaucheux breech-loaders in 1873.

SAUVEUR, Henri Mathieu. Gunsmith, 2 Soeur de Hasque, Liege and then from 1888 84 Rue Sur la Fontaine. Issued two patents in 1908.

SAUVEUR, Henri Joseph Marie. Gunsmith, 84 Rue Sur la Fontaine, Liege, with his father and then in 1908 he moved to 627 Rue St Leonard, Liege. He issued two patents in 1914.

SAUVEUR, Victor. Gunsmith, 33 Rue Laixhaut, Herstal. Issued a patent in 1912.

SCHOLBERG et GADET. Gunsmiths, 22 Rue Jonfosse, Liege. Registered at the Liege proof house 1861-85. Awarded six patents between 1870-73. Trade mark is initials S&G.

**S&G**

SCHOLBERG, Guillaume. Gunsmith, 22 Rue Jonfosse, Liege. Registered at the proof house between 1885-90. Given ten patents in Belgium.

SIMONIS, Albert. Gunsmith, 18-20 Rue Trappe, Esneux, Liege. Registered at the Liege proof house from 1873 to 1903. Issued nineteen patents between 1876-94 several of which were specific to Lefaucheux revolvers and many others applicable to all revolvers. In 1890 Simonis opened the Liege Gunsmiths College. Between 1893-1921 he registered at least ten trade marked names for his revolvers, 'The Unique', 'The British Boxer', 'Imperator' and 'Triumvir' for example. Trade marks star over AS and A&S over crossed muskets.

☆
**A S**

A⨯S (A&S over crossed muskets)

SIMONIS (FABRIQUE D'ARMES REUNIES S.A.). Gunsmiths, 31 Rue Jonfosse, Liege, from 1903 then in 1921 26-28 Rue Charles Morren. This is the partnership of Albert Simonis, Anton Bertrand et Fils, Riga et Cie, and Pirlot put together to manufacture the very popular semi automatic pistol the 'Dictator'. The trade mark was F.A.R and F.A.R crowned.

**F.A.R**

♛
**F.A.R**

SPORCQ, Eugene. Gunsmith, 1 Rue Coupe, Mons. Registered at the Liege proof house from 1861-80. Awarded six patents including improvements to Lefaucheux arms.

**SPORCQ**, Charles. Gunsmith, and ammunition manufacturer. Main address is 12-16 Grand Rue, Mons, which was in operation 1904-42. From 1912 they had another works at 8 Rue Anglais, Liege and a retail outlet at 3 Rue Ste Aldegonde as well as an ammunition factory on the Rue Etienne. The trade mark was a circled CS.

Detail from a 1930 advert for Charles Sporcq. After 90 years the venerable pinfire is still popular.
*littlegun.be*

**TANNER**, Thomas et Cie. Registered 1857-71. Known to have embellished pinfire pistols.

**WARNANT-FRANSQUET**, Leonard. Son of Leonard Joseph, married to Marie Fransquet. Gunsmith, Hoignee, Cheratte. Awarded six patents between 1859-87, including number 8592 of 1860 for improvements to Lefaucheux rifles and other weapons in collaboration with his father.

**WARNANT**, Emile. Gunsmith, 99 Rue Lamark, Liege. Active on the Liege proof house list between 1905-34. Awarded four Belgian patents. Trade mark is EW starred.

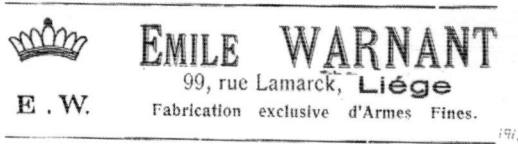

Emile Warnant address block.

**WATHELET**, A. Gunsmith, 115 Rue St Leonard, Liege. Awarded a patent in 1860 for an iron cartridge for Lefaucheux type weapons.

## UNKNOWN MANUFACTURERS

Unfortunately due to time and two world wars the details of some manufacturers have been lost. This 7mm revolver is marked with initials AD crowned. The usual way to register your mark was to use your initials, Christian name first, unless someone else had already used them then they would be reversed. This could possibly be for Antoine Defourney, c.1894-1929 or more likely Alexandre Dupont 1868-91 but unfortunately we will probably never know.

♛

A D

*Author's collection*

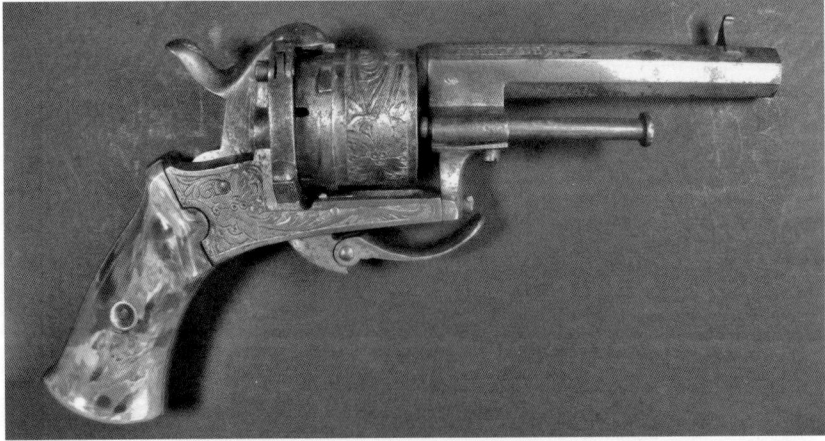

A rather flamboyant 7mm with deeply engraved floral decoration on the frame, barrel and cylinder; the malachite green grip is made from resin. To finish the job the revolver is nickel plated. It is marked with the initials VP which could be for Vivario Plomdeur as his dates are 1836-70 and the inspector's mark is a crowned Z for 1853-77, but I have been unable to see any of his work to look for similarities.

V P

*Author's collection*

**ORPHANS.** It was the practice in the Belgian gun trade to form groups of small manufacturers in Societe Anonym or anonymous societies, to enable them to bid for large orders which they could not handle individually. These societies could be in existence for only a few weeks or months and therefore did not waste money registering trade marks, which leaves us with thousands of unmarked, unattributable weapons.

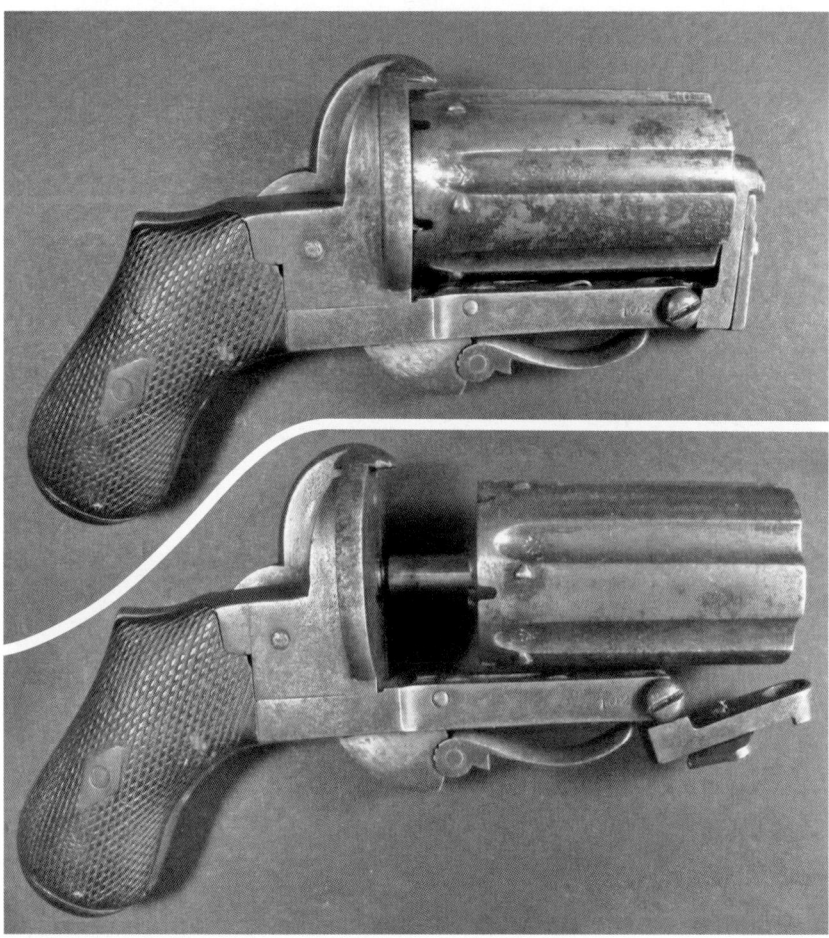

7mm double action pepper-pot (coup de poing, or fist punch in French) with a quick release latch on the front. Twist the latch to the right and the front bar drops on a hinge releasing the cylinder for loading; the cylinder pin is used to eject any spent cartridges. The cylinder has Birmingham proof marks but the gun itself was almost certainly made in Belgium.

*Author's collection*

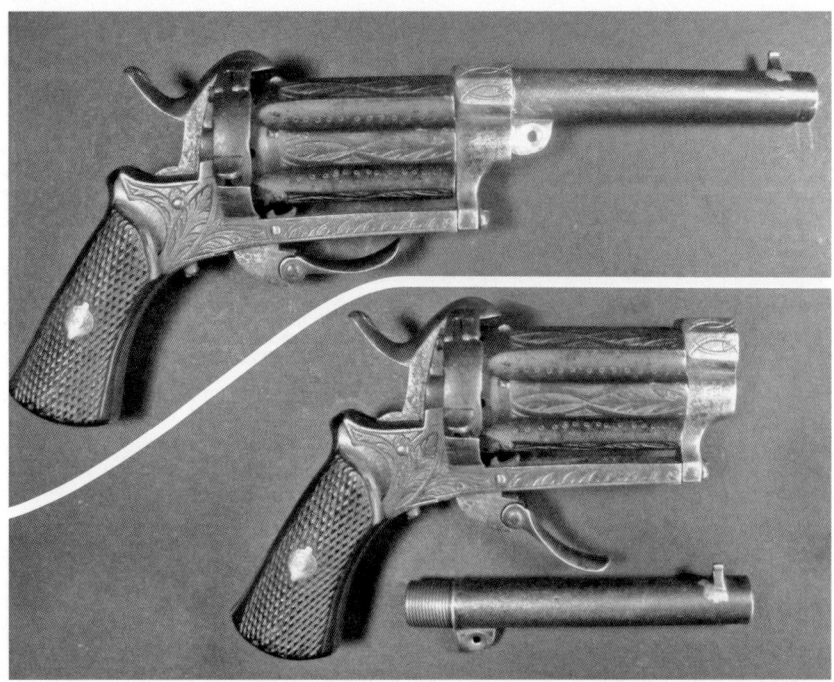

Liege proofed 7mm pepper-pot revolver with screw on barrel. Fine foliate engraving and circle and dot punch work covers the cylinder and frame, even down to the grip screw head. The rifled barrel is 3½ inches long. *Author's collection*

A snub nosed, Liege proofed pepper-pot revolver. This is basically a standard 7mm revolver with the barrel assembly replaced by an hourglass metal fixture which is screwed into the end of the frame and holds the cylinder pin in place.
*Author's collection*

Entry level 5mm revolver, even without an ejector rod and in poor condition it is worthwhile having it in a collection just to show an example of the smallest calibre regularly available. This example lacks any engraving or embellishment. However, it does have a rifled barrel and cross-hatched grips. The proof and inspector's marks date this to *c*.1895-*c*.1910.

*Author's collection*

A nickel plated 5mm pinfire revolver in its purple silk lined pipe case. The barrel measures only 55mm, with an overall length of 130mm. Dating to the last decade of the 19th century this example lacks rifling, but has finely cross-hatched grips. *Author's collection*

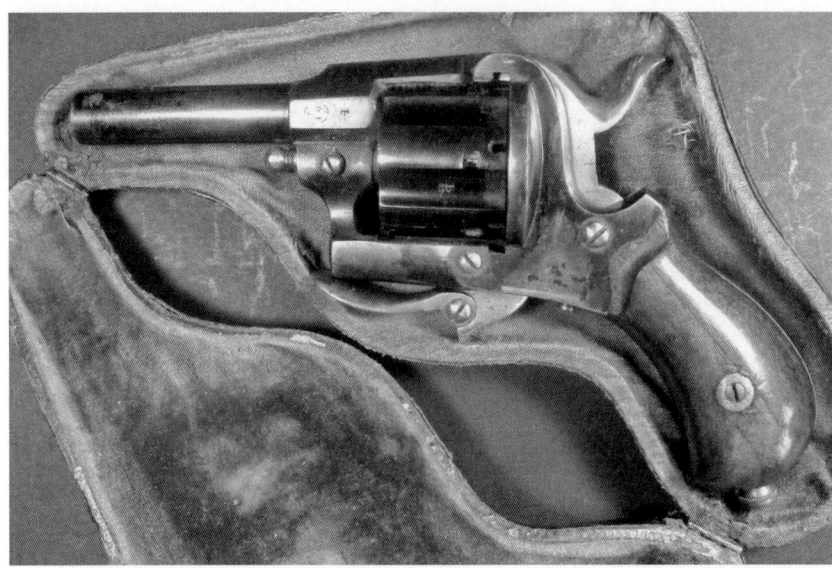

A top quality solid frame 7mm revolver complete with its travelling pipe case. The full fluted cylinder and birds head grip date this weapon to the 1870s. The L-shaped spring provides a firm open or closed position for the loading gate. Retaining much of its original finish and with strong springs this is a really nice addition to any collection.

*Author's collection*

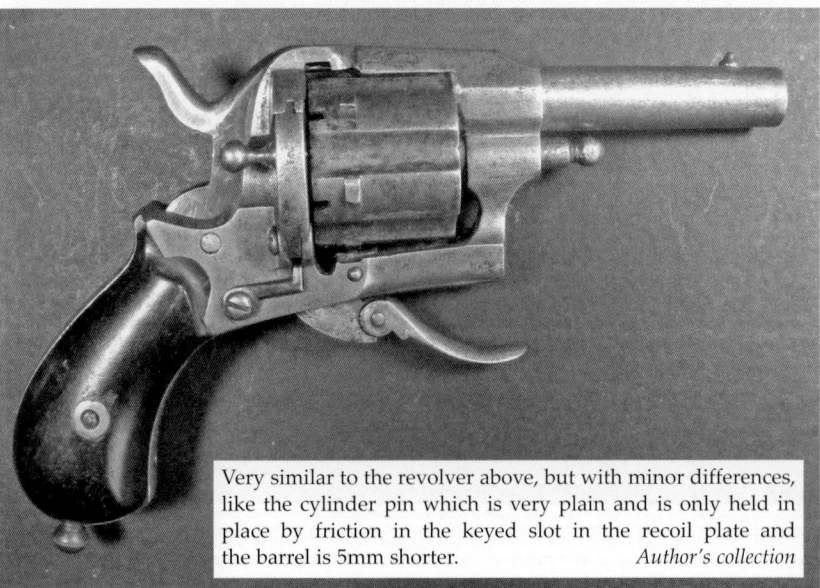

Very similar to the revolver above, but with minor differences, like the cylinder pin which is very plain and is only held in place by friction in the keyed slot in the recoil plate and the barrel is 5mm shorter. *Author's collection*

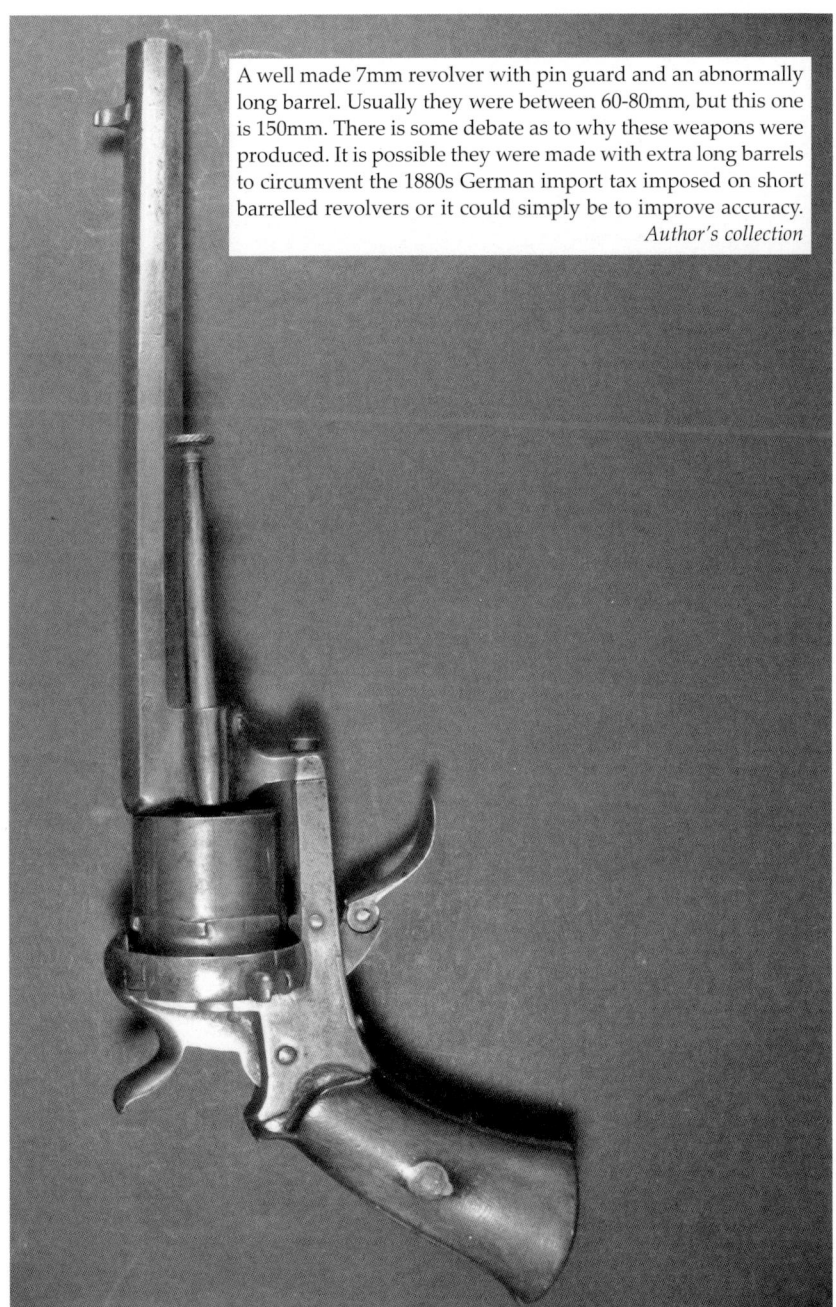

A well made 7mm revolver with pin guard and an abnormally long barrel. Usually they were between 60-80mm, but this one is 150mm. There is some debate as to why these weapons were produced. It is possible they were made with extra long barrels to circumvent the 1880s German import tax imposed on short barrelled revolvers or it could simply be to improve accuracy.
*Author's collection*

This 12mm revolver is marked with a tiny TB on the barrel just above the inspector's crown Z mark. This is almost certainly the barrel maker's mark, not the final manufacturer. Dating from the late 1860s or early 1870s this is a fine example of the gun maker's trade.
*Author's collection*

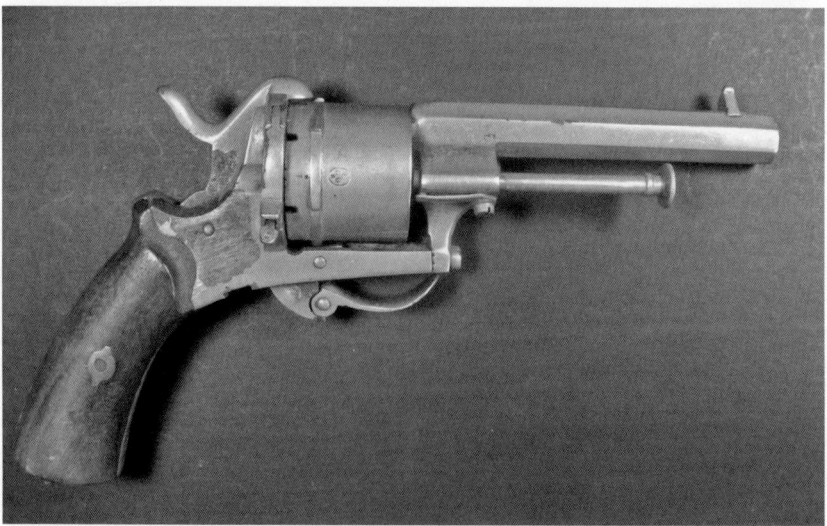

Dating to between 1877 and 1893, this nickel plated six shot 7mm revolver is unmarked apart from the proof and inspector's stamps. There is a tiny stared R which could be one of the worker's marks.
*Author's collection*

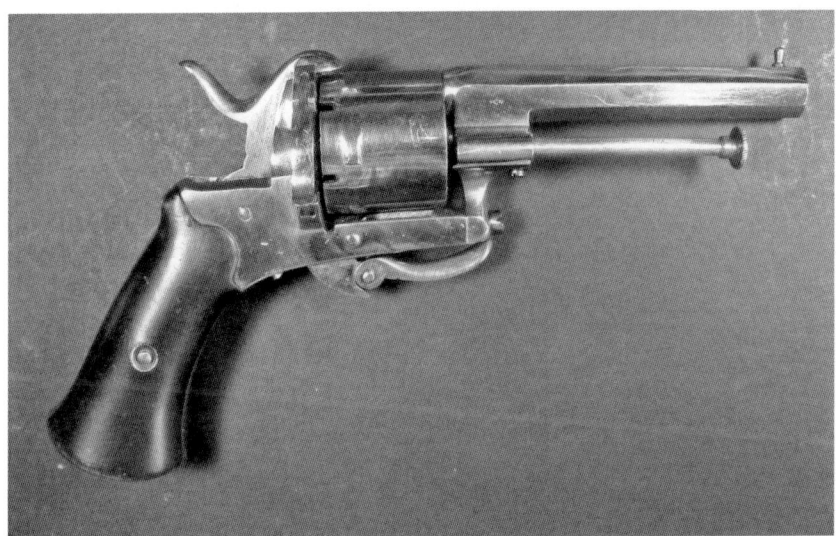

A finely made and well plated single and double action 7mm revolver with rifled barrel and ebonised grips. Why on earth they did not think this worthy of a maker's mark I cannot tell. Dating to the 1880s.  *Author's collection*

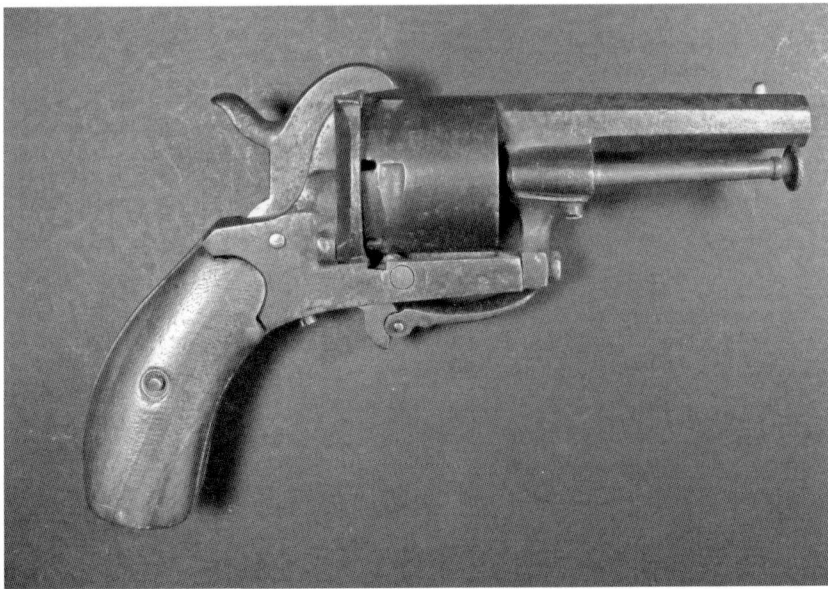

Very much in the style of Charles Clement with his distinctive pull down loading gate this revolver is unmarked except for the useful dating aid of 1913 stamped next to the inspector's mark.  *Author's collection*

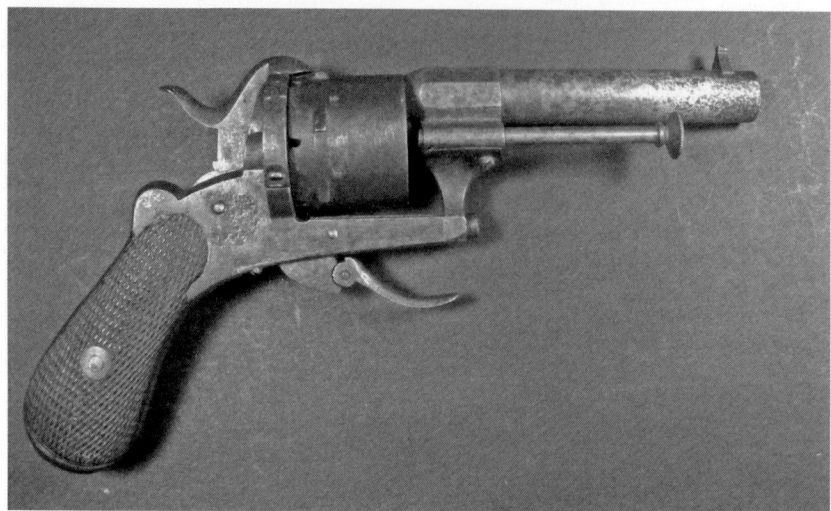

This Birmingham proofed 7mm revolver was almost certainly made in Liege and then sent across to England for finishing and testing. In 1859 the United Belgian Fire Arms Manufacturers opened a depot in London for just this purpose. This particular gun dates to the mid 1860s.
*Author's collection*

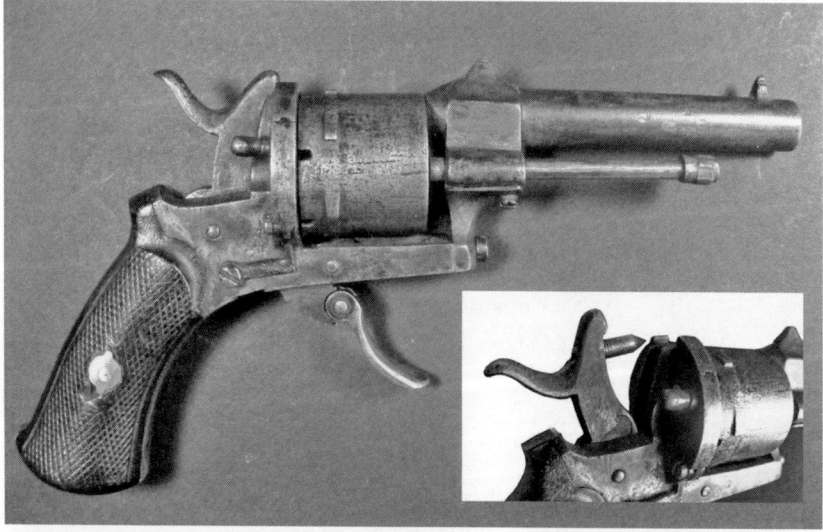

A 6 shot 7mm revolver, apart from the Liege proof and inspector's mark it carries no other markings. It has obviously had some care spent on its design and manufacture with the rear sight on the barrel lug rather than a simple V cut into the top of the hammer, loading gate spring and ebonised and cross-hatched grips. It dates to the late 1860s but at some later date has been converted to accept centre fire ammunition.
*Author's collection*

# European Pinfire Manufacturers and Patent Holders

Dating to the 1870s this 9mm revolver operates in both single and double action. Set up for military use it has a heavy frame, longer than usual rifled barrel and a lanyard ring on the butt.

*Author's collection*

A fairly standard layout 9mm revolver of *c*.1880 aimed at officers requiring a personal side arm when they were either still being issued percussion weapons or none at all. No markings on the weapon, not even on the butt strap which is where individual craftsmen would mark the pieces they had a hand in to claim their pay.

*Author's collection*

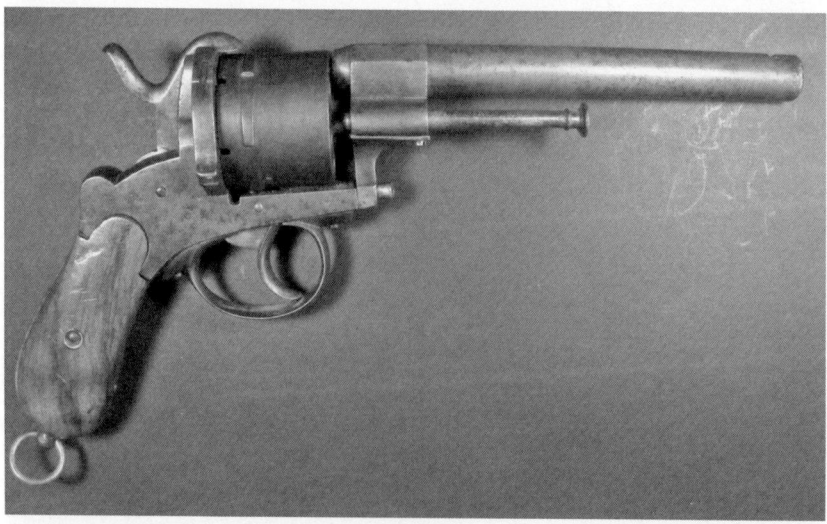

This 12mm revolver has Birmingham proof marks, but was almost certainly sent over to England in the white to be finished and tested over here. French gunsmiths would send their weapons to Liege to be proofed because it had a good reputation, but the English still preferred to see a Birmingham mark which was deemed to be the best.

*Author's collection*

A top quality, but except for the finely cross-hatched grips, very plainly finished 12mm revolver. This workmanlike finish was supposed to attract the military man or adventurer. Missing its lanyard ring this gun is otherwise in superb condition.    *Author's collection*

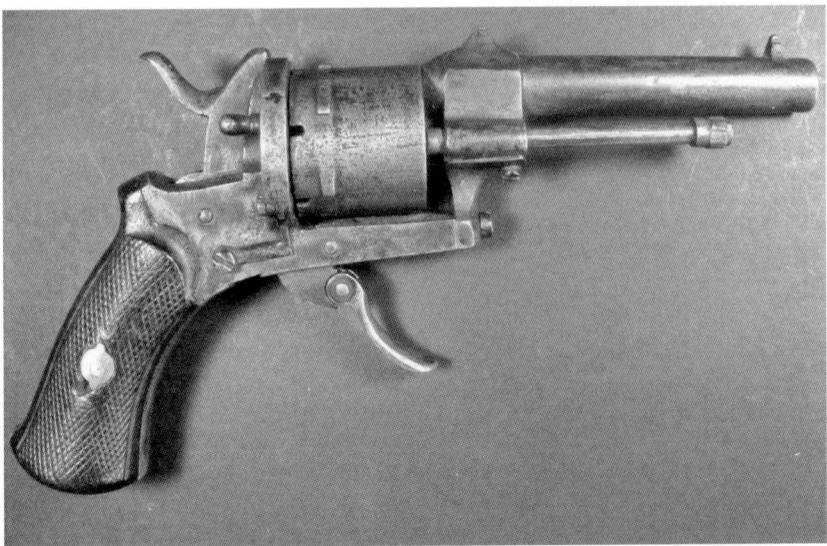

Dating from the mid 1870s this 12mm bears no marks other than the Liege proof and inspector's marks. It does have the rugged good looks and no doubt cheap price that would attract an impoverished Lieutenant looking to equip hiself.

*Author's collection*

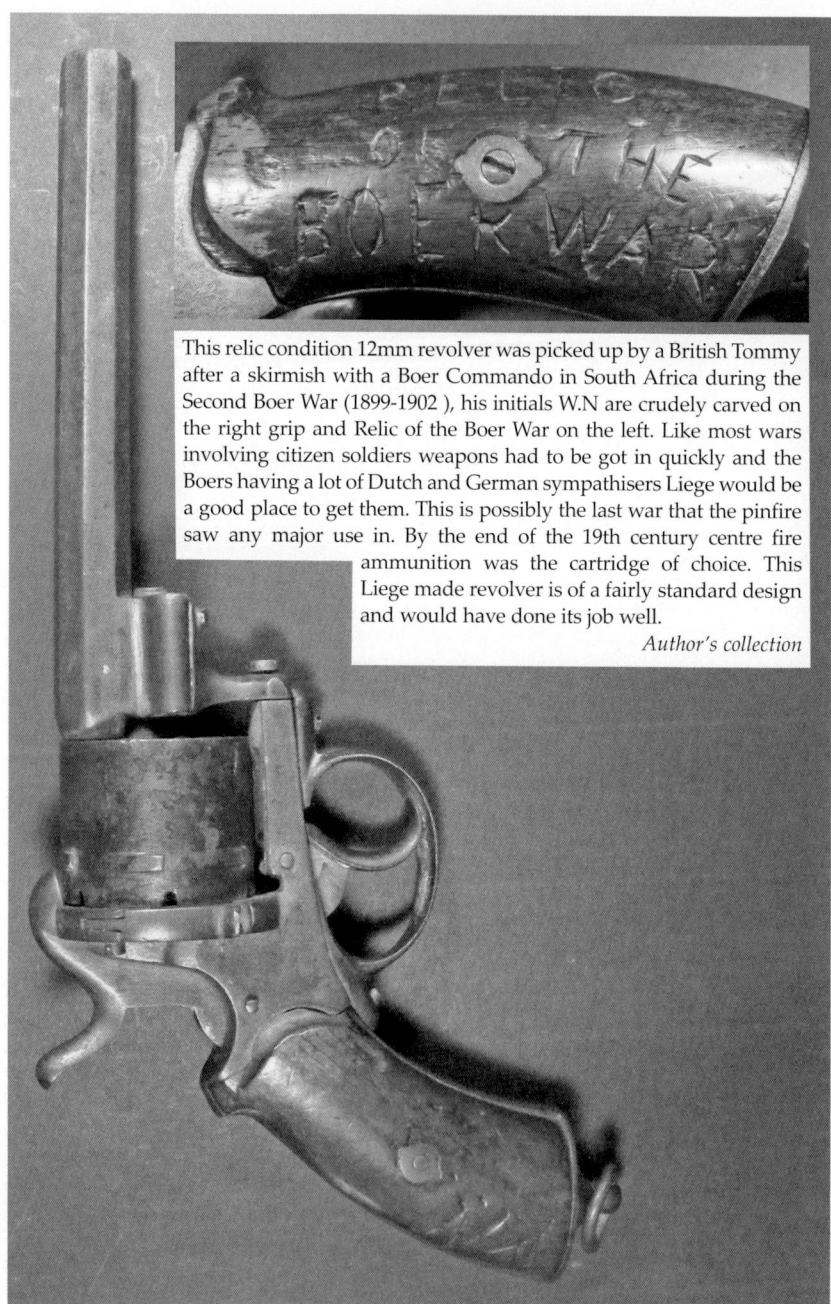

This relic condition 12mm revolver was picked up by a British Tommy after a skirmish with a Boer Commando in South Africa during the Second Boer War (1899-1902 ), his initials W.N are crudely carved on the right grip and Relic of the Boer War on the left. Like most wars involving citizen soldiers weapons had to be got in quickly and the Boers having a lot of Dutch and German sympathisers Liege would be a good place to get them. This is possibly the last war that the pinfire saw any major use in. By the end of the 19th century centre fire ammunition was the cartridge of choice. This Liege made revolver is of a fairly standard design and would have done its job well.

*Author's collection*

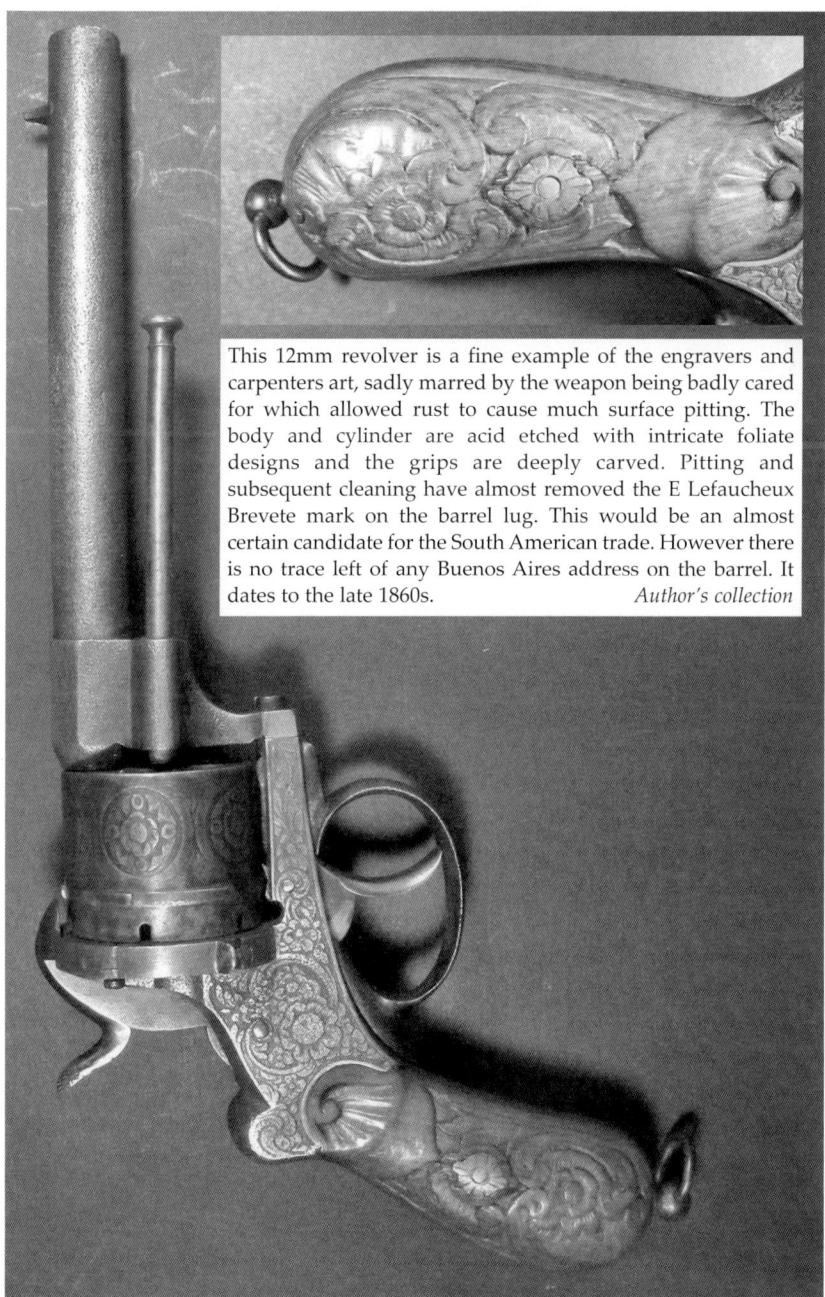

This 12mm revolver is a fine example of the engravers and carpenters art, sadly marred by the weapon being badly cared for which allowed rust to cause much surface pitting. The body and cylinder are acid etched with intricate foliate designs and the grips are deeply carved. Pitting and subsequent cleaning have almost removed the E Lefaucheux Brevete mark on the barrel lug. This would be an almost certain candidate for the South American trade. However there is no trace left of any Buenos Aires address on the barrel. It dates to the late 1860s. *Author's collection*

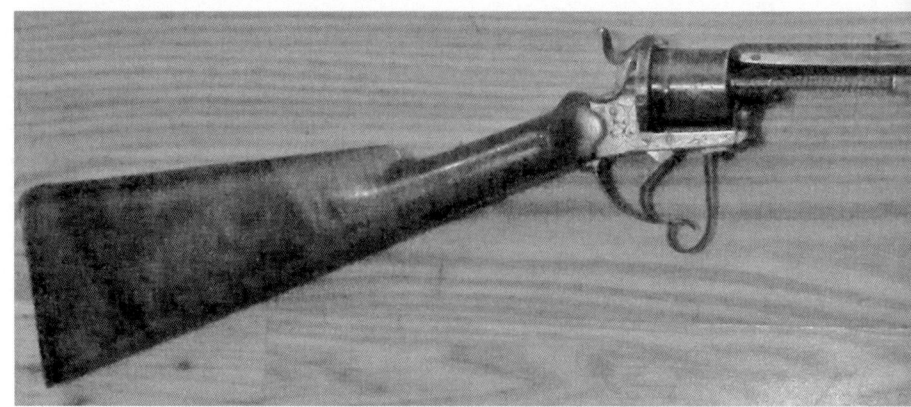

# L. ANCION-MARX
WAFFENFABRIKANT
26-28, Rue Grandgagnage, LÜTTICH (Belgien).

Ancion-Marx address block from the 1900-01 German catalogue.  *littlegun.be*

A fine workmanlike military style 12mm revolver, except for the smooth bore barrel which will have a serious effect on the accuracy of any shot of more than ten feet or so. Dating to the 1870s it is unmarked except for the standard proof and inspector's marks.

*Author's collection*

The ultimate pinfire revolver. was introduced by Eugene Lefaucheux in 1867 at the Paris Exhibition. The revolving rifle quickly became very popular and was copied by many Liege gunsmiths. It was available in 7mm, 9mm, 12mm and the huge 15mm calibres, this one being a 12mm example. The overall length of the rifle is 106cm and the barrel is 65cm. The sprung ejector rod and ornate trigger guard are standard for all manufacturers, as is the quality of the walnut stock.  *Author's collection*

## MANUFACTURE D'ARMES DE GUERRE
### DE CHASSE & D'EXPORTATION

 Jules PIRE & C<sup>ie</sup>

Jules Pire & Co address block from their 1899 catalogue. Manufacturers of arms for War, Hunting and Export.  *littlegun.be*

## FRANCE

**BARGERONT**, Auguste et Fils of Marseille. Awarded a French patent, number 78002, on 15th October, 1867 for a pinfire revolver with a hinged barrel which drops down to release the cylinder for easy loading and unloading by use of the cylinder pin. This invention was further protected by a Belgian patent, number 24146 on 9th September, 1868.

**BERJAT**, Paul. Gunsmith, 1 Place Villeboeuf. St Eteinne. Active *c.*1860–1879. Mostly known for his double barrel boot pistols.

**BERNY**, Edmond. Gunsmith, 46 Rue Poissonniere, Paris and 4 Rue Bassenge, Paris. Active 1866-72. His price list for 1866 shows the retail prices for 7mm Lefaucheux revolvers,

| Walnut grip | In the white | Satin finish | 10.00 |
|---|---|---|---|
| Walnut grip | Polished finish | Satin finish | 11.00 |
| Walnut grip | Four polishes | Satin finish | 11.25 |
| Ebony grip | Four polishes | Satin finish | 11.75 |
| Ebony grip | Nice engraving | Hollow polish | 13.75 |
| Ivory grip | Nice engraving | Hollow polish | 20.00 |
| Ivory grip | Strengthened barrel | Polished white | 18.50 |

**BERTONNET**. Gunsmith, 56 Passage Choiseul, Paris. Active between 1840-60. Bertonnet also had a retail outlet at 50 Calle St Martin, Buenos Aires between 1867-71 where he would send his guns 'in the white' for finishing and engraving.

**BOITARD**, P. Gunsmith, 6 Grande Rue St Jacques, St Etienne, then 4 Grande Rue St Roch.

**BRUN**. Gunsmith and engraver, Paris. Manufactured high quality pinfire revolvers which were richly engraved and guilded.

**CHAMELOT, J et DELVIGNE**, Henri Gustave. Chamelot was a gunsmith from Liege who went into partnership with Henri Delvigne, an Army officer and Legion of Honour recipient of German extraction who lived in Paris. Between the two of them they were awarded seven Belgian patents and two additions for pinfire revolvers from 1862 to 1865 before moving on to gain a further six for centre fire revolvers before the death of Delvigne in 1876.

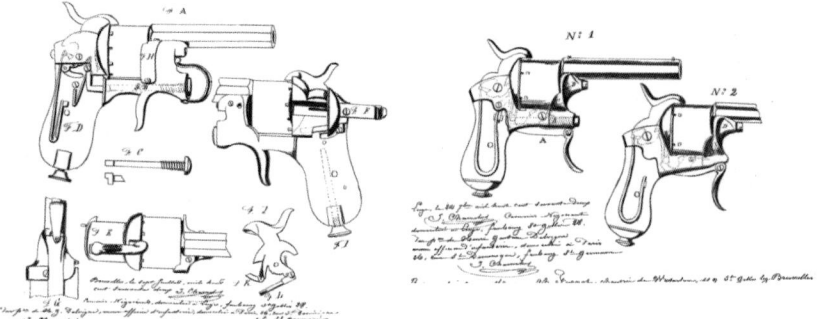

Variant 1, patent 12861 of July 1862.
*Author's collection*

Variant 2, patent 13241 of October 1862.
*Author's collection*

1st variant, patent 12861 of July 1862 shows a small revolver with an apparently split frame held together with a wedge and screws, a side hammer action and a loading gate on the left-hand side which was pulled back on its sliding bar let into the frame. The mechanism was borrowed from Chaineux's patent of 1858.

The 2nd variant, patent 13241 of October 1862 has a one piece frame and a similar layout to the first patent.

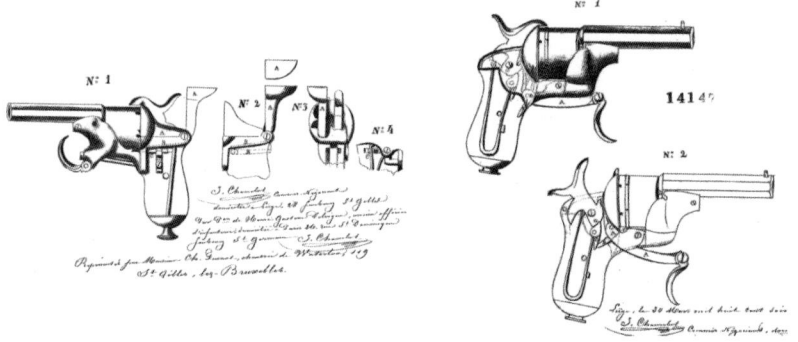

Variant 3, patent 13258 of December 1862.
*Author's collection*

Variant 5 patent 14147 of 30th March, 1863.
*Author's collection*

3rd variant, patent 13258 of December 1862 has the loading gate changed to a much simpler sprung, pull back version that was to become standard.

The 4th variant was entered in early 1863 as an addition to patent 13258. In this variant the weapon is cocked by a bar attached to the trigger which runs into the butt. When the trigger is pulled the bar is pushed to the rear and this forces an L shaped lever upwards to rotate the cylinder and cock the gun.

5th variant, patent number 14147 of March 1863. In this version the bar has been replaced by a long arced cam which does away with the L shaped lever as the new cam runs directly from the trigger to the hammer and cylinder ratchet.

The 6th variant is an addition to patent 14147 and is a double action, spurless hammer version of variant 5.

Variant 7, patent number 14955 of 5th September, 1863, shows the removal of the ejector rod from the butt and now fitted to the left side of the barrel and running along a guide rod.

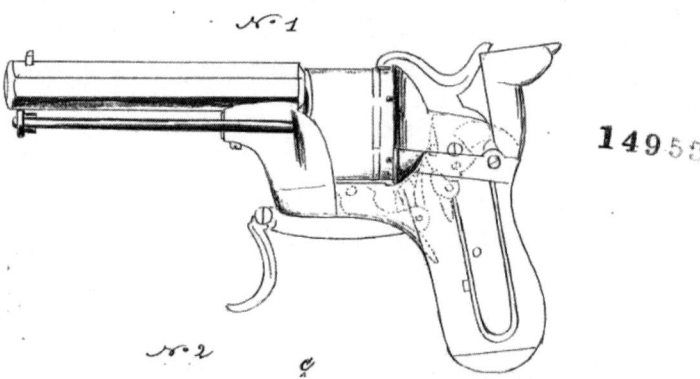

Variant 7, patent 14955 of 5th September, 1863.     *Author's collection*

Model 8. All the previous variants were unnamed or numbered. However, the next patent was described as Model 8. Awarded patent number 16152 on 5th May, 1864 this version proved to be by far the most popular. In this version when the trigger is pulled back one end of the rocker cam cocks the weapon and spins the cylinder while the other lifts up to catch the protrusions on the cylinder stopping it moving once it is in place.

Model 9, patent number 18901 is very similar to Model 8 but with the addition of a pin shield.

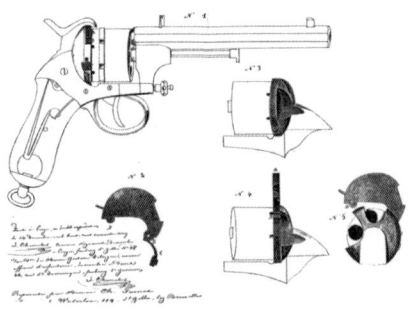

Model 8, patent 16152 of May 1864.　　　Model 9, patent 18901 of December 1865.
*Author's collection*　　　　　　　　　　　　　*Author's collection*

Chamelot and Delvigne Model 8. The lower photograph shows the spring loaded ejector rod and lift up loading gate.
*Author's collection*

**CLAUDIN**, Ferdinand. Gunsmith and retailer, 38 Boulevard des Italiens, Paris. Active 1861 to 1878. Awarded patents for several innovations including this pin guard seen on a Lefaucheux Paris made double action revolver.

*Left*: Details of the Claudin patent number 74 pin guard. *Centre*. The left hand side of the guard with the patent details. Top view the guard. Each side is screwed separately on to the recoil plate. *Above*: The address stamped on to the top of the barrel. A photograph of the complete gun can be seen in the French gunsmiths section under Eugene Lefaucheux. *Author's collection*

**DELHAXE**, Joseph. Gunsmith and cutler, 8 Rue Thevenot, Paris. Manufactured pinfire revolvers with bayonets attached. He was awarded seven Belgian patents between 1870 and 1871. It was normal at this time for French inventers to have their guns patented and manufactured in Liege as it was so much cheaper than France. The first patent, number 27820, was issued on 24th April, 1870 for an improvement on the Dolne-Bar patent for combination guns. Patent 27818 was issued on 28th June, 1870 for a pepper-box revolver called the Sortie de Bal or Out to the Ball gun. Patent 27819, issued on the same day was for a combination knuckleduster, revolver and knife weapon called the Casse Tete or Headache / Puzzle gun. The fourth patent was for an improvement on the Flobert system. Patent 28697 was for improvements to his own patent 27819. Patent 28909 was again for improvements to the Flobert system. His final patent, number 29717 of 5th December, 1871 was for improvements to his Flobert system patent 28909.

**DEVISME**. Gunsmith, 36 Boulevard des Italians from 1840 to 1860 and then 12 Rue Helder, Paris.

**DRIVON-OZONNE**. Gunsmith, Paris. His patent number 66991 of 20th December 1865 is for the quick ejection of spent cartridges. With the weapon on half cock the barrel and cylinder can be tipped down on a hinge fitted to the frame. As the barrel drops the top 5mm of the cylinder is pushed out by a cam and the six empty cases are ejected as one.

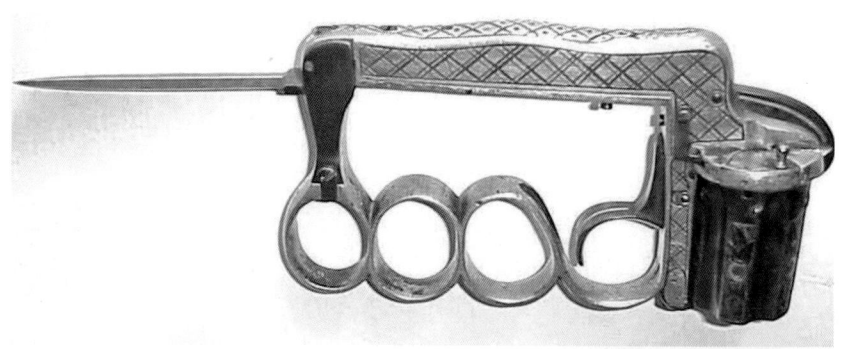

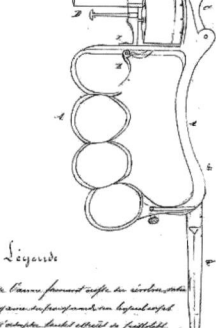

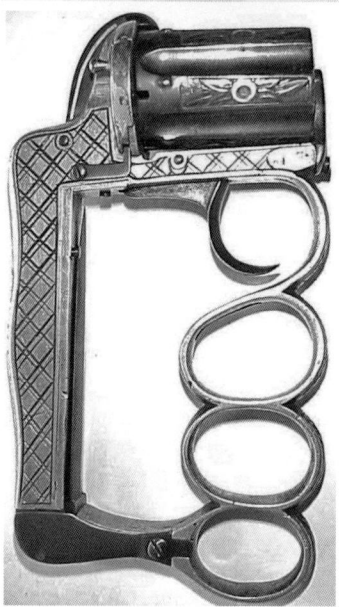

Delhaxe 'Puzzle Gun'. Patent drawing and two views of the gun in open and closed positions.
*Horst Held*

**DUMONTHIER**, Joseph Celestine. Gunsmith, Faubourg St Martin and Rue Petits Hotel, Paris. As well as the Paris sites he also had a manufacturing plant at Sailleville in the Oise region of France. Dumonthier is well known for making pistols with bayonets attached and walking cane shotguns which he patented in 1876. Active in the gun trade between c.1840 and c.1890.

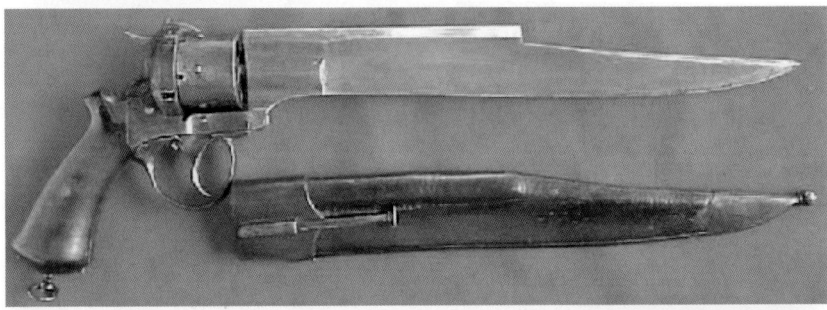

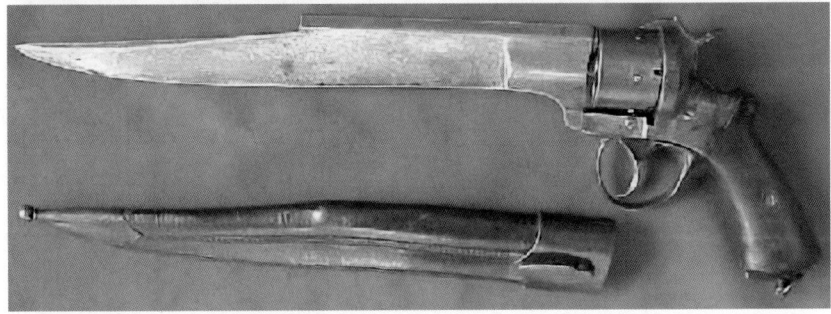

Dumonthier knife revolver with a Bowie blade. Note how the ejector rod is fitted to the scabbard so that when the blade is sheathed it is in the correct position to be used normally.
*Horst Held*

**ESCOFFIER**, Felix. Director of civil gun making at the St Etienne state armoury. Produced pinfire revolvers in the style of Javelle.

**EYRAUD**, Jean Baptiste. Gunsmith, Rue Etienne, St Etienne. Awarded a patent on 15th November, 1858, for a side hammered revolver design. His innovation was to design a catch which when released allowed the barrel to swing out to the left allowing it to be removed for easy loading and unloading. He was active in St Etienne between 1825 and 1860.

**FAURE-Le PAGE**, Gunsmith, established in 1716 in Rue Baillif, Paris and later 8 Rue de Richelieu, Paris. During the riots in Paris of 1830 rather than have their works ransacked they handed out all their guns to the rioters. Later, after they had stopped making arms and had gone into the leather trade they moved to number 21 where they are still trading.

Faure-Le Page's well stocked gun shop at 8 Rue de Richelieu, Paris in 1920. The signs above the windows tell us they were makers of arms for hunting and deluxe weapons.
*Christian Feron*

**FEMINIER**, Albert. Gunsmith, 96 Grand Rue, Ales. Active *c*.1870s.

**GALAND**, Charles Francois. Gunsmith, 296 Rue Vivegnis, Liege, and 13 Rue Hautville, Paris. He also had a shop at 3 Rue Richer, Paris.

**GOUERY-CANAT**, JF et Cie. Gunsmith, Paris. Made pinfire copies of Col LeMat's patent revolver. These had 7mm calibre nine shot cylinders. However, what should have been a larger calibre central barrel was a dummy.

**GUYOT**. N. Gunsmith, 8 Rue de Lyon, Paris. Active 1880 to 1890.

**GUERRIERO**, A. Gunsmith, Paris.

**HOULLIER-BLANCHARD**, Charles. Gunsmith, St Etienne, 1845-55, then 36-38 Rue de Clery, Paris, 1855-62 followed by a move to 4 Rue Clery, Paris from 1862 until 1872. On the death of Charles the works were taken over by his son Jacques, and bought out in 1872 by Mr CH Pidault who moved to 25 Rue Royal, Paris.

An 1895 advertisement for Houllier & Blanchard. *Christian Feron*

**JARRE**, J et Cie. Gunsmiths, 28 Boulevard Poissioniere, Paris, and 81 Rue Lafayette, Paris. Between 1859 and 1862 he was awarded several patents for his famous 'Harmonica' pistols.

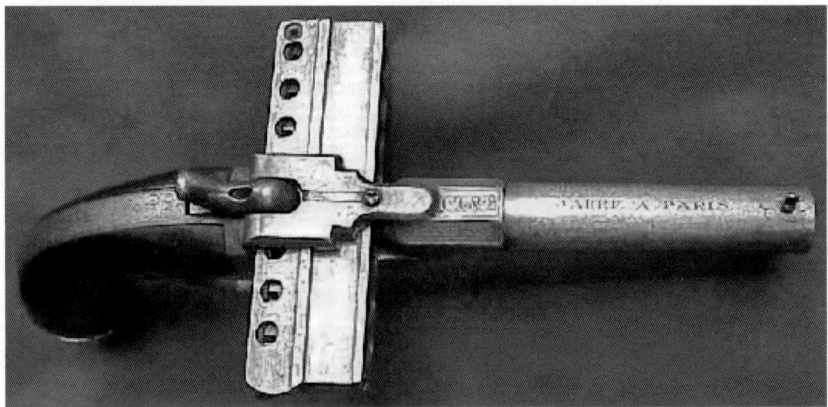

Jarre 'Harmonica' eight shot pistol. As the trigger is pulled a claw moves the magazine left to right lining up the pin for the next shot. *Horst Held*

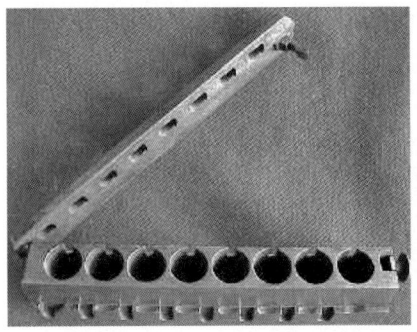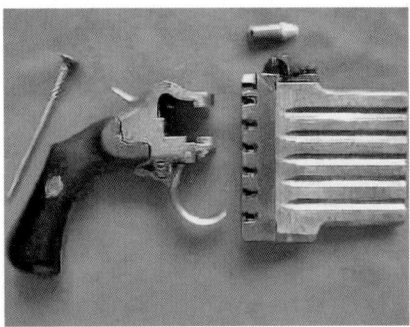

Magazine for the pistol bottom previous page, showing the method of securing the cartridges, and. a six barrel version Jarre pepper-pot with the chambers acting as barrels.

*Horst Held*

**JAVELLE**, M. Gunsmith, St Etienne. Patented quick cylinder release. As the fancy bar beneath the barrel is pulled to the right, jaws holding the end of the cylinder pin part and release the barrel and allowing it to drop down on a hinge, The cylinder can then be pulled off the cylinder pin.

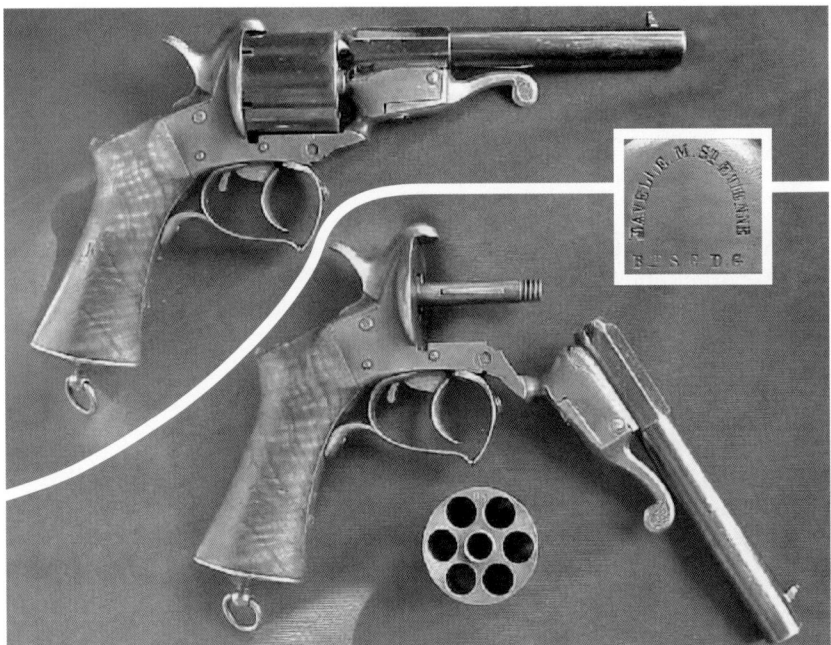

Javelle patent 9mm revolver shown in the firing position and the reloading mode. Javelle's mark is also shown.

*Horst Held*

**LAINE**, Jean Baptiste. Gunsmith, 21 Rue Rivoli, Paris. Active c.1860 to c.1880.

**LAURENT**, Edmond. Gunsmith, 20 Rue des Changes, Toulouse. Active from 1876 to 1900. Laurent's price list for July 1878 shows Lefaucheux revolvers at the following prices.

| | | | | |
|---|---|---|---|---|
| 7mm | Liege steel | Walnut grips | Polished | 6.75 |
| 7mm | Liege steel | Ebony grips | English engraving | 8.50 |
| 7mm | Liege steel | Ebony grips | Richly engraved | 10.50 |
| 7mm | Cast steel | Ivory grips | Domed cylinder head, safety catch | 15.00 |
| 9mm | Liege steel | Walnut grips | Polished | 9.50 |
| 9mm | Liege steel | Ebony grips | Richly engraved | 12.00 |
| 9mm | Cast steel | Ebony grips | Domed cylinder head, safety catch | 17.50 |
| 12mm | Liege steel | Walnut grips | Polished | 11.50 |
| 12mm | Liege steel | Ebony grips | Polished and richly engraved | 15.00 |
| 12mm | Cast steel | Ebony grips | Domed cylinder head, safety catch | 21.00 |

**LEFAUCHEUX**, Casimir. Gunsmith, 37 Rue Vivienne, Paris. See chapter one for detail.

**LEFAUCHEUX**, Eugene. Gunsmith, 104 Rue Lafayette, Paris. See chapter one for detail.

**Le PAGE**. Gunsmith, 22 Rue Bourg-L'Abbe, Paris. Active 1823 to 1857.

**Le PAGE FRERES**, sons of Le Page senior above. Gunsmiths, 12 Rue D'enghien, Paris. Active 1857-70, then renamed Le Page et Chauvot until 1880, then renamed Chauvot-Lepine-Piot-Le Page until 1887. In 1887 a further change in the partners happened and they became Piot-Le Page et Lepine until 1890 when the final change of name to Piot-Le Page occurred. The company ceased trading in 1980.

**LEROUX et LEROUX**. Gunsmiths, 31 Rue Richelieu, Paris. Pinfire patents awarded in 1851, 1852 and 1861.

A 6 shot 7mm Eugene Lefaucheux Model 1856 self-cocking pinfire revolver, octagonal barrel, the breech stamped with 'LF' mark and number 8884, and engraved 'FD. Claudin Brevete a Paris'. Engraved overall with scrolls, the pin guard stamped 'Claudin Brevete', with folding trigger stamped 'Invr E.Lefaucheux Brevete'.   *Author's collection*

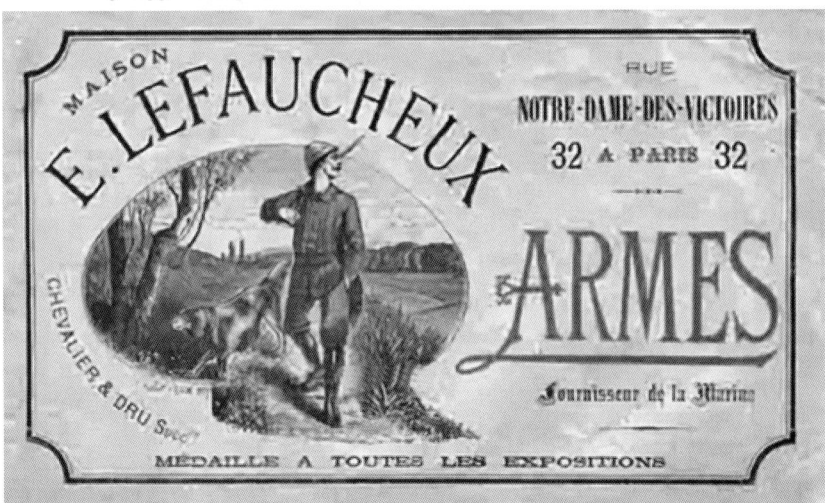

Front cover of Eugene Lefaucheux's post-1881 catalogue. Note the name Chevalier & Dru who took over in that year.   *Cristian Feron*

**LOEVEN**, L. Gunsmith, 63 Rue de Rivoli, Paris. Active c.1850.

**LORON**, Pierre Antoine. Gunsmith, 24 Rue des Bons Enfants, St Etienne.

**MERIEUX**, Michel Napoleon Isadore. Gunsmith, 9 Place d'Armes, Poitiers. Active c.1850-80.

**ROLLAND et RENAULT**. Gunsmiths, 3 Boulevard St Martin, Paris.

**SALLES**, E. Gunsmith, Beziers.

**SIGAUD-BARNERIAS**, Jacques. Gunsmith/cutlers, Thiers, Arance. Converted ready made pinfire revolvers by fitting sprung blades and bayonets.

**VOYTIER**, F. Gunsmith and retailer, St Etienne.

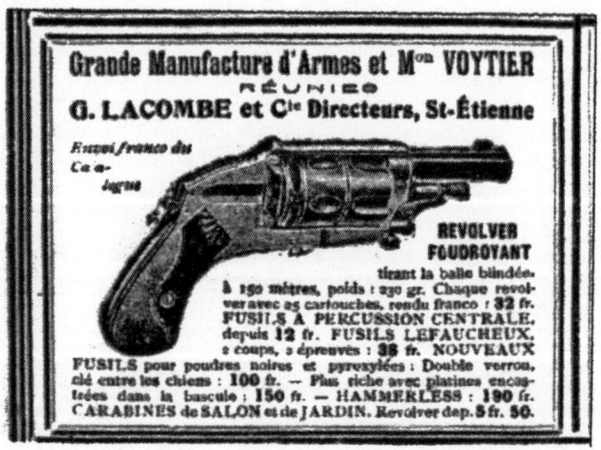

c.1895 newspaper advert for Voytier of St Etienne.

**WINDISCH**, Jean Louis. Retailer, 6 Place de L'ville, Nimes. Active c.1834 to 1881.

**ZAOUE**, Georges. Gunsmith, Place Royale, Marseilles. Active c.1841-70. Manufactured multi shot pinfire revolvers with up to 20 shots, also 15mm revolving rifles.

Apart from the serial number 2966 this French 9mm revolver, c.1860, is totally unmarked. It is, however, remarkable in three ways. Firstly the cannon mouth barrel, secondly the unusual retaining screw for the ejector rod and finally the exaggerated bag grip.

*Author's collection*

## GERMANY and AUSTRIA

**BAADER**, Bernhard & Sohn. Gunsmith, Munich, Germany. Active from 1854 to c.1880.

**CHRISTIANS**. Cutlers and gunsmiths, Solingen, Germany. From 1863 some edged weapon manufactures in Solingen began making pinfire revolvers to compete with the arms factories of Liege. However, by the 1880s they were unable to match prices and so production more or less ceased. Christians continued on in their major business as cutlers until the 1970s when they closed.

**DRGM**. Not a manufacturers mark but DEUTSCHES REICH GEBRAUCHTS MUSTER, or German Empire Utility Model, sometimes seen on later pinfire revolvers.

**DRPA**. Another official mark DEUTSCHES REICH PATENT ANGEMELDET, meaning German Empire Patent Applied For and sometimes seen on later pinfires.

**EBBEKE**, Bernhard. Gunsmith, Herzberg, Germany. Active c.1860s.

**EBERT, Paul & Sohn.** Gunsmiths, Germany. Founded 1822 and by 1876 the son had taken over.

**FRANKENAU**, O. Gunsmith, Germany. Issued patents in the USA and Great Britain for his purse pistol.

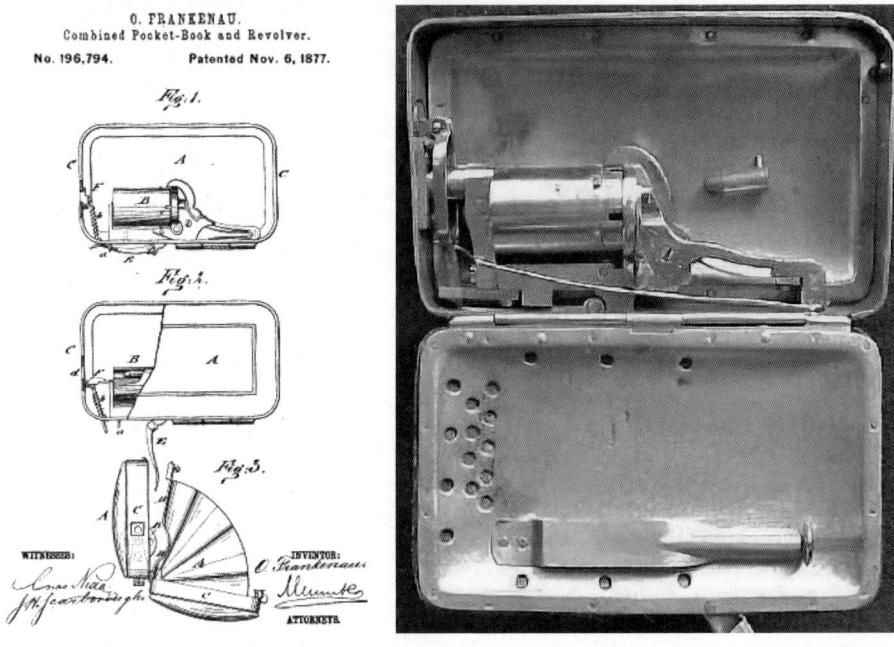

Frankenau's purse pistol patent of November 1877. On the right the interior of the purse showing the 5mm revolver.  *Horst Held*

**FUKERT**, Johann. Gunsmith, Weipert, Austria. Born 1811, died 1900. Active 1830 until 1870, father of Gustav.

**FUKERT**, Gustav. Gunsmith, Weipert, Austria. Active pinfire manufacturer 1870 to the late 1880s. His luxury weapons were sold to most of the Royal houses in Europe. Notably he was Gunsmith to Emperor Franz Joseph the first of Austria. He was awarded nine patents in Austria, Germany, Switzerland and Great Britain.

**HALLANG**, Emile **& BUCHNER**, Benno. Gunsmiths, Suhl, Germany. Founded 1862, still active in the late 1870s.

# European Pinfire Manufacturers and Patent Holders

**HASSER**, L. Gunsmith. Bad Swabach, Germany.

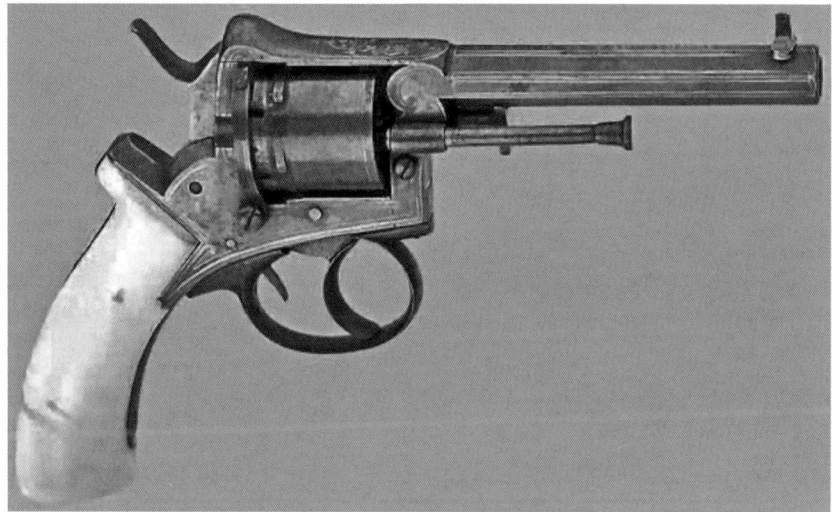

9mm engraved pinfire revolver with Ivory grips by Hasser of Bad Swabach.   *Horst Held*

**HOHMANN**, F and R, Cutlers and gunsmiths, Solingen, Germany. Active c.1860 to c.1880. One of their sidelines was the production of shaped sugar products.

**HOLLER**, A E & Co. Cutlers and gunsmiths. Solingen, Germany. As well as their shop in Solingen they had outlets in Paris and London. Their products included weapons, iron, steel and cutlery.

Trade adverts for Hohman and Holler from the 1873 Handbook of German Industry.   *HBoGI*

**KIRSCHBAUM.** Cutler and gunsmith, Solingen, Germany. Active from the early 1860s until 1883 when they merged with Weyersberg & Co.

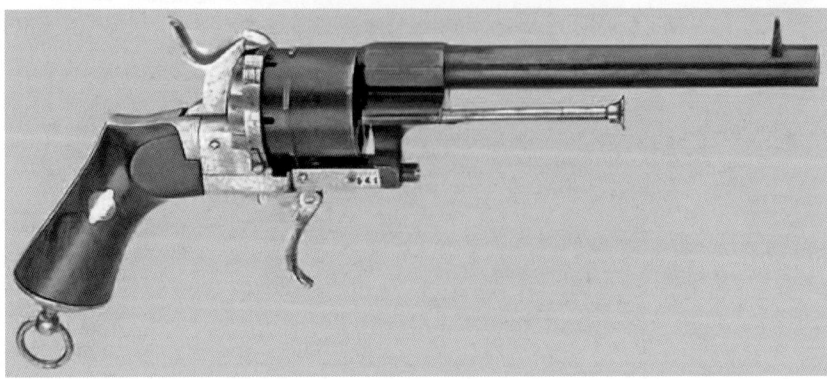

7mm pinfire revolver by Kirschbaum of Solingen c.1860. *Horst Held*

**KNAAK**, Georg. Wholesaler and retailer, 212 Friedrichstrasse, Berlin, Germany.

**LANGENHAN**, Frederich. Gunsmith, Zella, near Suhl, Germany. Company founded in 1842. Makers of luxury weapons, revolvers and pistols, still active in 1887.

**LEUE**, Heinrich & **TIMPE**, Johan. Gunsmiths, 79 Friedrichstrasse, Berlin, Germany. Luxury weapons from 1863 until the 1880s.

Advertisements for Knaak, Langenhan and Leue & Timpe, Berlin.
*HBoGI*

**LUNESCHLOSS.** P D. Cutler and Gunsmith, Solingen, Germany. Active gunsmith 1860s-*c*.1880s. 'Exporter of revolvers to all transatlantic countries' Trade mark PDL on all revolers.

Advert for P. D Luneschloss of Solingen.
*HBoGI*

**MEHLES,** Hippolit. Wholesaler and retailer, 61 Auguststrasse, Berlin, 1878's to 1879, then 160 Walter Friedrichstrasse, Berlin, Germany until 1885 when the shop closed. In 1893 Hippolit emigrated to America.

Advertisement for Hippolite Mehles gun emporium. Some unmarked guns were stamped with his own trade mark of an eagle over H M over BERLIN.
*Author's collection*

**MILLER**, Franz Zavier, BADER & Sohn. Gunsmiths, Germany. Active c.1830-c.1870.

**PETERLONGO**, Johann. Gunsmith, Innsbruck, Austria. Maker of cavalry and infantry officer's revolvers, 1854 to 1918.

**RIFLEMANN**, H. Cutler and gunsmith, Solingen, Germany. Active gunsmith c.1860 to c.1880.

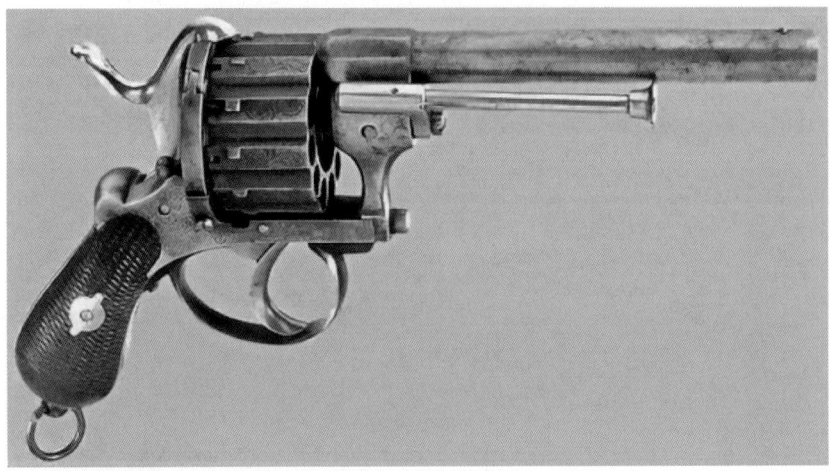

12 shot 7mm revolver by Riflemann of Solingen.          *Horst Held*

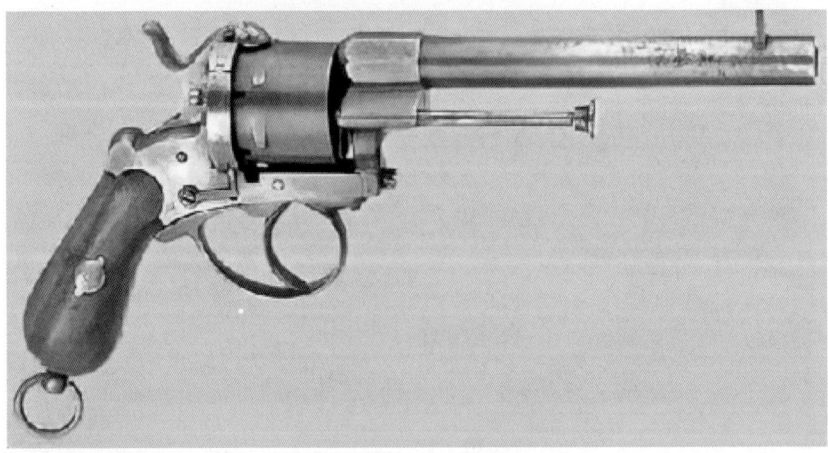

9mm revolver also by Riflemann of Solingen.          *Horst Held*

**ROCH, STEIR & Co.** Gunsmiths, Suhl, Germany. Founded 1865.

**ROOS,** Ulrich, & Sohn. Gunsmiths, Stuttgart, Germany. Active *c.*1840 to *c.*1870.

**SCHILLING,** Valentin Christoph. Gunsmith, Suhl, Germany. Active 1849 to 1900.

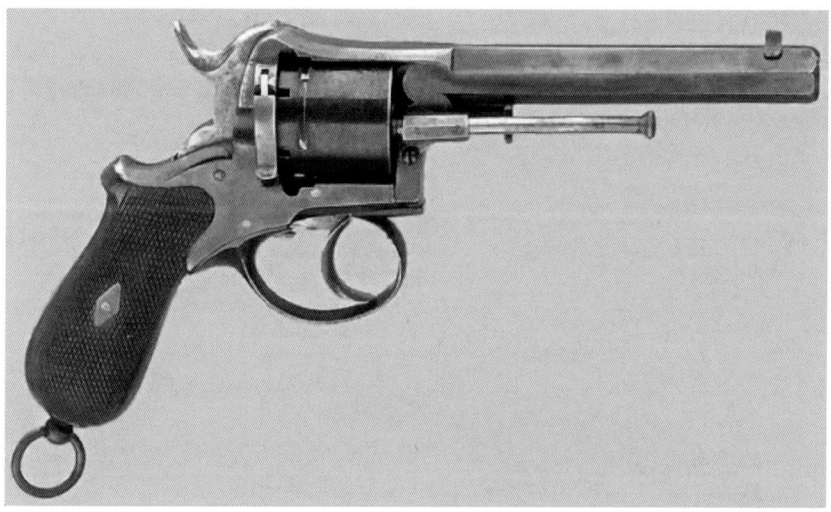

9mm solid frame revolver by Schilling of Solingen, *c.*1870.   *Horst Held*

**SCHMIDT & HABERMAN**. Gunsmiths, Suhl, Germany. Founded 1858 and still active in 1872.

**SCHULER.** August. Gunsmith, Suhl, Germany. Active *c.*1870 to *c.*1880.

Trade adverts for Schmidt & Habermann and H. Schuler.   *HBoGI*

**SPANGENBERG**, Ferdinand **& SAUR**, Johan-Paul. Gunsmiths, Suhl, Germany. Founded in 1838 when the two arms companies merged. In 1849 the company grew with the addition of Sturm becoming Spangenberg, Saur & Sturm. In 1873 the name changed again to Saur & Sohn. They are still in the arms business as SIG-Saur.

1873 advertisement for Spangenberg & Saur. *HBoGI*

Spangenberg & Saur 7mm revolver with a patent quick barrel release fitted. *Inset*: Detail from the revolver showing the quick release bar lifted to remove the barrel. *both Horst Held*

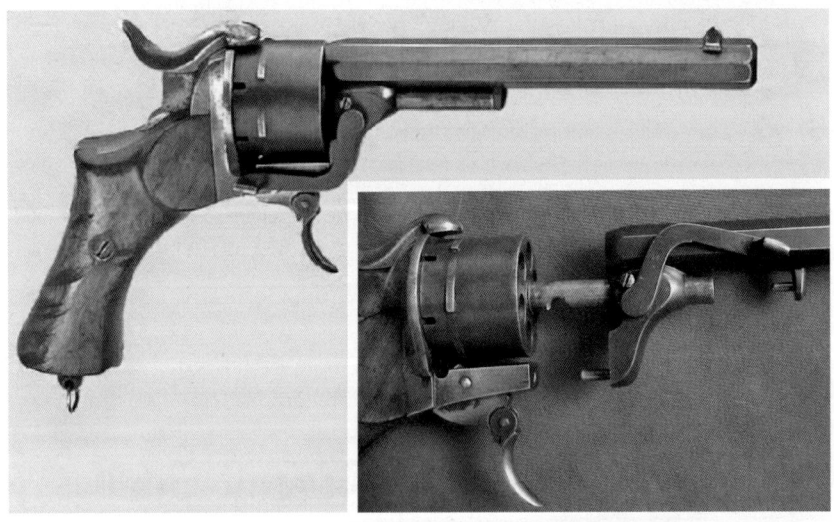

**SWARZ & FELZ**. Gunsmiths, Mersch, Luxembourg. Maker of luxury weapons, active *c.*1860 to *c.*1890.

**THILON**, Moriz. Retailer, 3 Goldschmied Gasse, Vienna, Austria. His advertisement from 1872 shows the price in French Francs of both Lefaucheux shotguns and revolvers in regular and better grades.

**TIROLER WAFFENFABRIK PETERLONGO, RICHARD MAHRHOLDT & Sohn.** Gunsmiths, Innsbruck, Austria. This is the successor company to Peterlongo & Sohn. In 1898 Richard Mahrholdt was appointed manager of the Peterlongo works and stayed with them until 1918 when he left to form his own company Mahrholdt & Sohn. Peterlingo limped on under Johann's son until 1939 when he purchased it and absorbed it into his own company. Richard died in 1949 and his son became manager.

**WEYERSBERG, KIRSCHBAUM & Co.** Cutlers and gunsmiths, Solingen, Germany. Founded in 1883 with the merger of Kirschbaum and Gebruder Weyerberg, two of Solingen's biggest metal work companies. Now known as WKC they are still trading in the edged weapon business, including the takeover of Wilkinson Sword's ceremonial sword business.

German Orphan 7mm revolver, marked with the double crown U proof mark.
*Author's collection*

# ITALY

**ADAMI**, Vincenzo. Gunsmith, Cosenza. Active c.1850 to c.1880. Produced very ornate pinfire revolvers as well as needle and percussion weapons and shotguns.

**GLISENTI**, Francesco and Isidoro. Iron founders and gunsmiths, Brescia. The brothers set up an iron foundry in 1859 and shortly afterwards began the production of firearms. In 1861 the Italian government ordered 5,000 Lefaucheux model 1858 to be produced in France by Lefaucheux and under licence in Italy by Glisenti. They were classified as the Pistola a rotazione da Carabineeri Reali Modello 1861. After the contract was fulfilled they continued to make civilian versions. Having their own mine, blast furnace, two hammer forging workshops and a machine shop they soon became a major supplier of arms. In 1907 all weapon production ceased and the company concentrated on cast iron work.

**GUERRIERO**, Alexandre Count de St Angelo. Engineer and weapon designer, Lombardy. The Count, unable to get his designs made in Italy moved to Paris and with his agent Eugene Breuer began production in Liege, Belgium. In his first patent, registered in Britain as patent number 628 of 5th March, 1863, the design split the cylinder when a lever was pushed down and the barrel was pulled forward. This allowed the used cases to be ejected and new ones loaded.

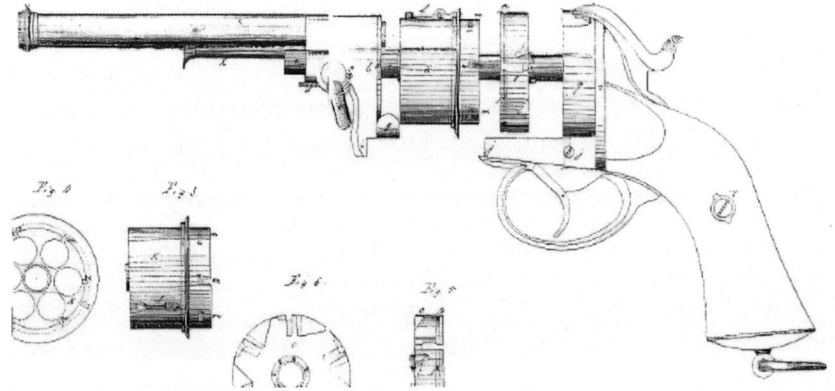

Guerriero's French patent, dated 30th January, 1864. *Author's collection*

Side view of pistol from Guerriero's first French patent. *Author's collection*

Guerriero's second French patent, number 71326 of 16th April, 1866  *Author's collection*

Systeme Italien 7mm pinfire revolver by Guerriero, produced in Liege. The inset photograph shows the cylinder in the open position after the (replacement) cylinder pin has been withdrawn. *Author's collection*

Guerriero's second patent was for a revolver with a swing out cylinder. The cylinder pin is moveable and can be withdrawn by sliding it along underneath the barrel. Once the pin is clear of the cylinder put the hammer on half cock and tip the weapon to the right. The cylinder falls out of the frame on a hinge and the recoil plate springs open ready to empty or load the cartridges. His design was patented in England in 1863 and France in 1866, but again the production was in Liege. The French Army began to show an interest in the weapon in January 1864 as they considered it better than the Model 1858 that the Navy had chosen. Under the representation of Gideon Marc, an officer of the 7th Bn Seine National Guard, it was presented to the Vincennes commission who put it on trial in 1864 and 1865. The Commission found that the weapon was prone to fouling and the mechanism was too fragile for military use and rejected it. In 1867 Guerrier showed his revolver at the Paris Universal Exhibition where it was the star of the show and many were produced and sold. The idea was later used by Smith and Wesson for their own revolvers.

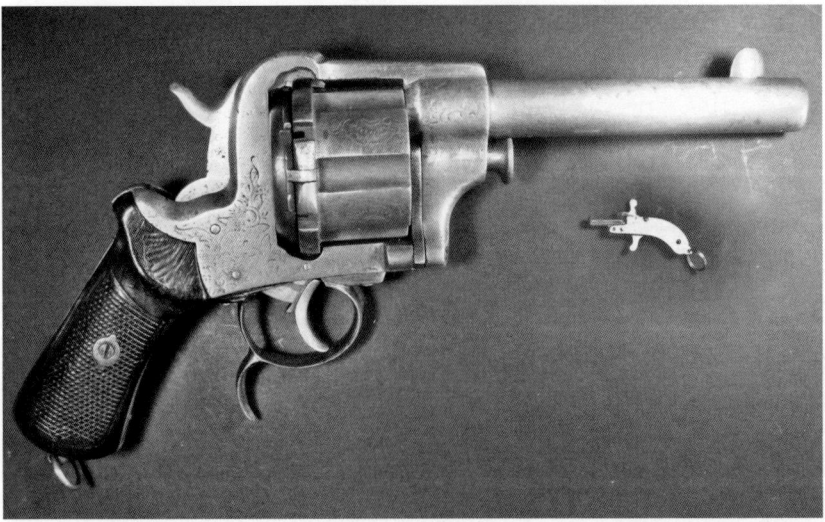

Little and large. This monster 15mm revolver is posed next to its tiny brother, the 2mm Mouse pistol, made in Liege to Breuer/Guerriero's French patent 71326 of April 1866. The barrel is 15½cm long and has a circumference of 7cm. Overall length is 32cm and it weighs around 1.2 kilos. The spur on the trigger guard is to assist in managing the fearsome recoil of this massive weapon. Overall this is a fine piece of engineering. The trade mark is G INV-BR Brevete for Guerriero inventor and patent holder, with crossed swords between. *Author's collection*

**MAZZOCCHI FRATELLI.** In the early 19th century Gaetano Mazzocchi was hired as the Keeper of the Papal Armoury with the exclusive right to make weapons for the Papal Army. This privilege was suspended during the Napoleonic Wars and then given to his sons Giovanni, Giuseppe, Pietro and Luigi on the resumption of Papal powers. Until 1850 the armoury was situated in Castel del Angelo in the Vatican but it had to be handed over to a French military mission and so they relocated to Rome proper. In their workshops they produced Artillery pieces and rifles for the Papal guards. In 1867 the committee in charge of approving handguns for the Papal Artillery had sight of a Chamelot and Delvigne revolver. They liked what they saw but asked for the design to be simplified. The new design was approved on 4th June, 1867 and on 21st February, 1868 Mazzocchi Fratelli were asked to manufacture the guns under licence. They were to make a total of 2,500 and provide them to the Papal Police at a rate of 40 per week. The brothers' link with the Pope was ended on 20th September, 1870 when the Risorgimento or unification of Italy saw the Papal powers reduced to the tiny enclave of the Vatican.

Mazzocchi Brothers 'Roman Pattern' or 1868 pattern revolver based on a modification to the Chamelot & Delvigne system. The serial number, 861 shows this particular revolver was made about 10th July, 1868 and the tiny stamp on the cylinder (AG) shows it was made by the craftsman Giovanni Amidoni. It also carries the mark of Fratelli Mazzocchi on the frame. *Author's collection*

MEROLLA FRATELLI. Gunsmiths, Naples. The Merolla Brothers Francesco, Giovanni and Salvatore were active in Naples from 1850 right through to 1920. The eldest brother, Salvatore, began the business in 1850 and he was joined by his brothers later in the 1860s. Salvatore was awarded Belgian patents in 1863 and 1864, but the patents issued from 1868 were under the Merolla Fratelli name. The three brothers continued to issue patents until the mid 1870s when Giovanni and Salvatore drop off the applications. Francesco claimed his last patent in 1920.

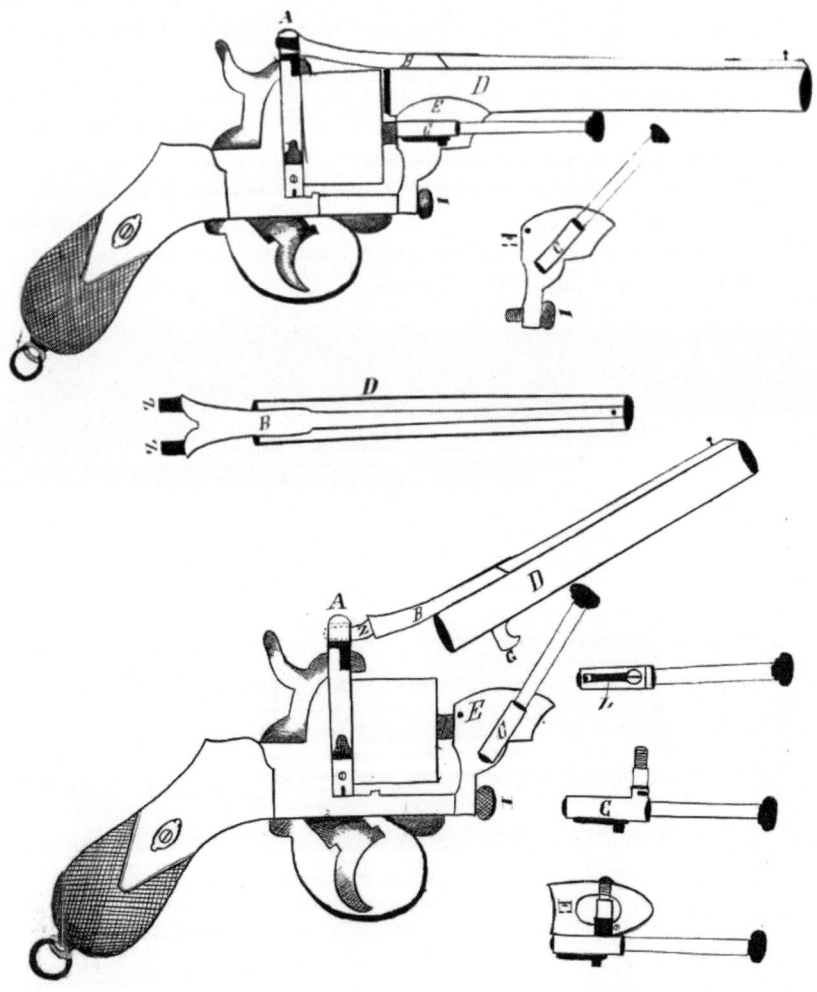

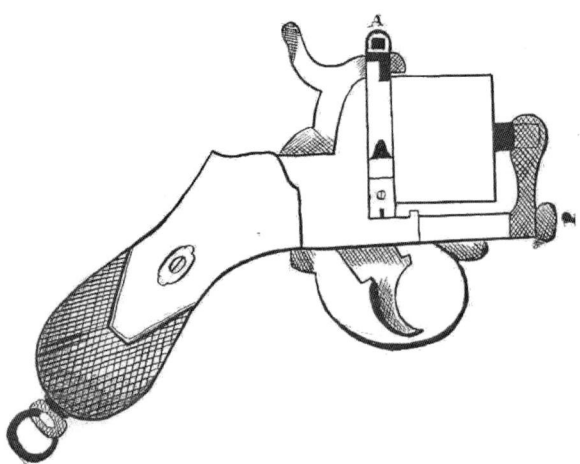

The two drawings on the facing page are details from Salvatore Merolla's patent number 21162 of 14th March, 1868. It shows a revolver with a top strap on the barrel that is hooked onto the top of the recoil plate. When the ejector rod is pushed upwards the barrel is released and can be removed. Removing a knurled nut holding the barrel release group allows it to be removed and replaced by a figure of 8 shaped fitting which turns the weapon into a pepper-pot as seen above.                                        *Author's collection*

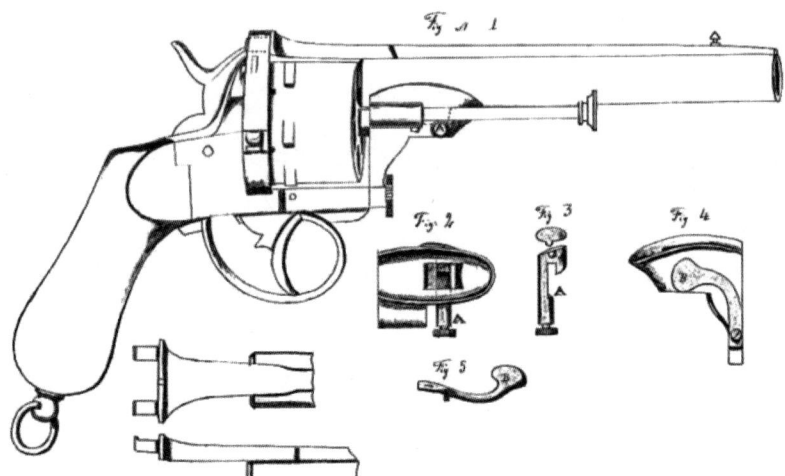

Merolla Fratelli patent from January 1868 again with a hooked on top strap. This time the whole barrel group comes off to allow the cylinder to be removed for ease of loading and unloading.                                        *Author's collection*

## SPAIN

*See also* Appendix One for other Eibar Gunsmiths.

**ARAMBURU**, Jesus. Gunsmith and cartridge maker. Vea se el Anuncia, Bilbao. Active *c*.1880.

**ARETIO**, Paulino. Gunsmith, 10 Caraiceria Viego, Bilbao. Active *c*.1860 to *c*.1890.

**ARETIO**. Gunsmith, Vitoria. Manufactured Lefaucheux double barrelled boot guns *c*.1860 to *c*.1880.

**ARIZMENDI Y GOENAGA**. Gunsmiths, Eibar. From 1881 to 1914. Manufactured revolvers until 1914 then automatics.

**ARRIABAN Y Cia**. Gunsmiths, Eibar. Active 1860 to 1870 manufacturing Lefaucheux Model 1858 under licence.

**ARTAMENDI**. Gunsmith, Eibar. Active 1860s producing double barrel pinfire pistols.

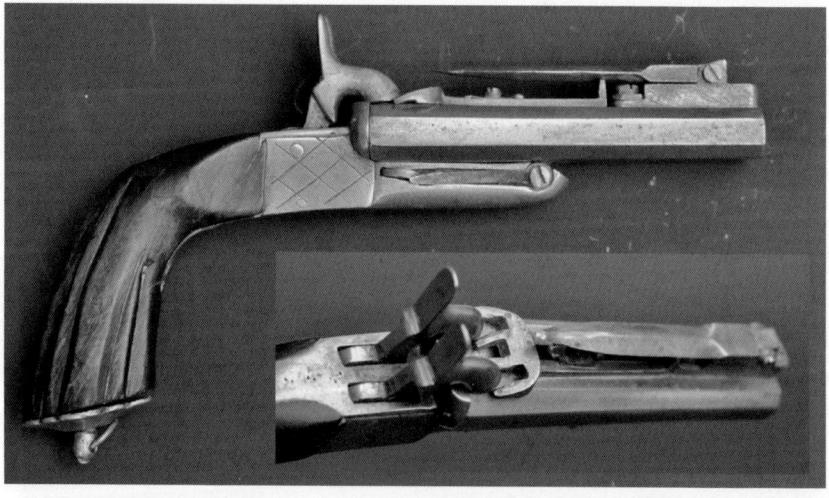

Unmarked double barrelled 12mm boot gun of the type made by Aretio in Vitoria and Artamendi in Eibar as well as by many other anonymous producers. This example has a fairly standard sprung blade at the front of the barrel, however, brazed to the top of the barrel is an unusual claw shaped sliding safety catch designed to prevent accdental discharge of the weapon when secreted about the body. *Author's collection*

**BARTHELET Y CHASTANG.** The Chastang brothers, Ernsto and Leopoldo, were employed by Orbea Hermanos to update the machinery at their factory because of their knowledge of the current Belgian and French methods. After they had completed the upgrade they decided to set up on their own but lacking the funds and contacts to be completely independent they formed an alliance with Barthelet who provided them with guns 'in the white' for finishing and decoration. They were one of the very few Spanish gunsmith to use a trade mark, Crown over B C, over the date of manufacture, circled. Active from *c.*1865 to 1875.

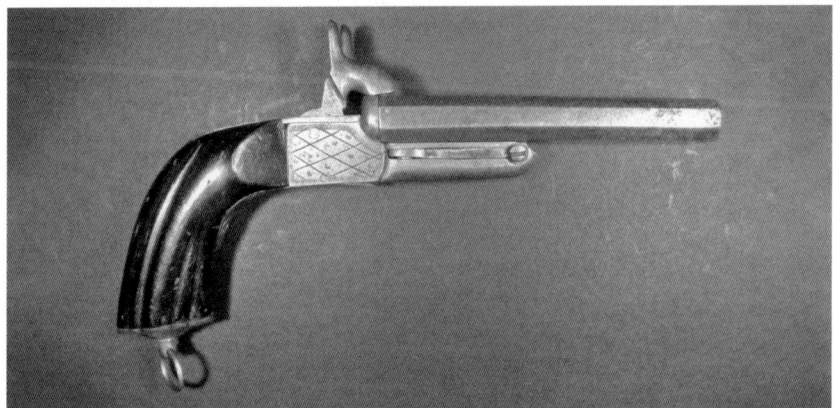

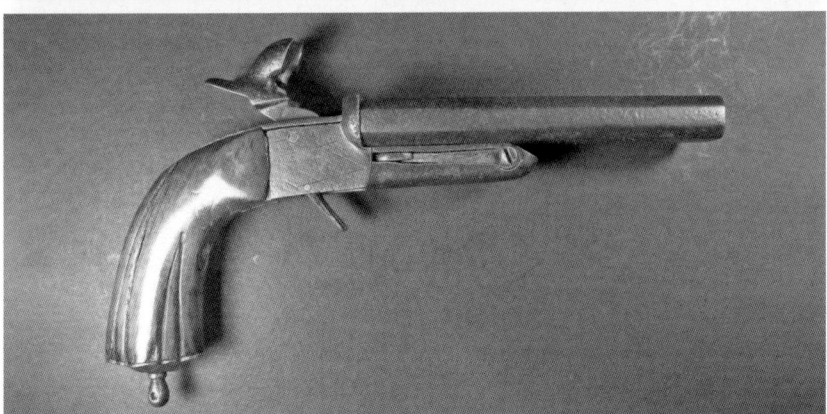

Two fairly standard 12mm pinfire boot guns as produced by the Spanish gun trade over the period from 1860 to 1910 without any change to the design. The smooth shape allows them to be safely slid into a boot without fear of a misfire. The triggers only become exposed as the hammers are pulled back. *Author's collection*

**BASCARAN**, Triaphon. Gunsmith, Eibar. Active 1881 to 1896 producing shotguns and pinfire revolvers.

**ECHANIZ**, Cristobel. Gunsmith, Eibar. Active c.1865 to 1914 then joined Arostegui (see other Eibar Gunsmiths) in partnership.

**GARATE, ANITUA Y CIA**. Gunsmiths, Eibar. Active as a workshop from 1893 to 1896 and a factory from 1897 to 1923.

**GUESALAGA**, Angel. Gunsmith, Eibar. Active c.1860 producing pinfire revolvers.

**GUISALAGA**, Pedro. Gunsmith, Eibar. Active 1850 to 1880 making Lefaucheux shotguns.

**JUARISTI**, Pablo. Gunsmith, Eibae. Active 1881 to 1914 producing pinfire boot pistols.

**ORBEA HERMANOS**. Gunsmiths, Eibar. The factory was founded in 1859 by the Orbea brothers Juan, Mateo and Casimiro. They began by copying Lefaucheux revolvers without a licence, but by 1863 this had been rectified. They were one of the largest firearms manufacturers in Spain until 1927 when the family-owned business split in two. The company was renamed Hijos de Orbia Cia. One half of the family took over the cartridge factory in Vitoria and the other remained in Eibar to continue the manufacture of weapons. By the late 1920s the factory had diversified into bicycle manufacturing. In 1936 arms manufacturing stopped as the bicycle business boomed.

The Orbea Hermanos factory in Eibar in 1910.
*Orbea.com*

Early 20th century advertisement.
*Orbea.com*

**OSA**, Andres. Gunsmith, Eibar. Active *c.*1880. Produced pinfire boot guns.

**ORTIZ de ZARATE**, Gunsmith, active 1881 to 1890.

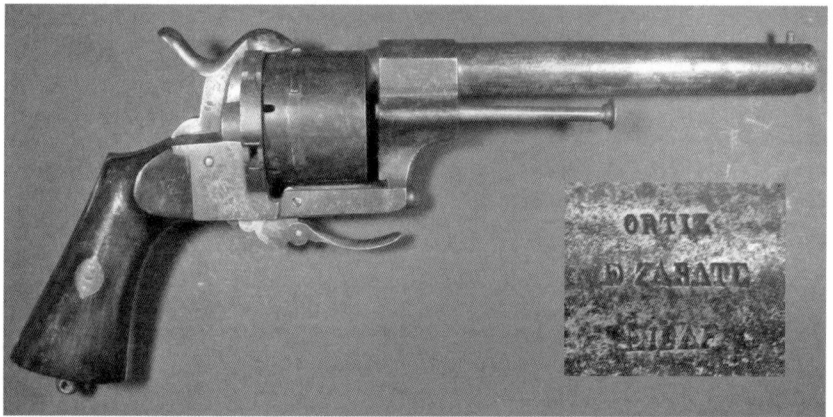

12mm pinfire revolver by Ortiz d Zarate of Eibar. It was quite common for Spanish revolvers to have fold-up triggers whatever the calibre of the weapon. *Author's collection*

**OVIEDO**. Spanish Royal Armoury. The first Royal armoury was set up up in the town Placencia de Armes, but from the 1860s the production of small arms was moved to the La Vega rifle factory in Eibar. The original factory was closed in 1880. The Lefaucheux Model 1858 was produced here for the Spanish National Guard and the Model 1863 for the Army. In 1865 the Royal monopoly for weapon manufacturing was broken and small armouries opened up like wildfire all over Eibar. By 1899 there were 1,149 registered gunsmiths out of a town population of 6,583.

**PAGUADA**, Francisco. Gunsmith, Eibar. Active making pinfire boot guns c.1858.

**TRUBIA**. Royal cannon factory. Opened in 1794 and still in production. Produced Lefaucheux Model 1854 revolvers for the Spanish Army.

Oviedo Royal Armoury produced Lefaucheux Model 1863 revolver. The overall pitting shows the poor quality of steel used in its manufacture, a failing of many Spanish-produced weapons.
*Author's collection*

Spanish orphan 12mm, 8 shot pinfire with pin guard and poor quality nickel plating.
*Author's collection*

# Chapter Four

## EVOLUTION OF THE PINFIRE CARTRIDGE

The first self-contained cartridge was patented in France on 29th September, 1812 by Samuel Pauly. The cartridge consisted of a screwed brass or copper cap which contained a primer. A central hole through the base of the cap and screw allowed the flame from a percussion cap to pass through to the main charge which was held in a paper or card tube which screwed onto the metal cap.

In 1816 Pauly modified his design with the addition of a striker pin which struck a fulminate cap. From 1812 Pauly manufactured rifles, shotguns and pistols using this system at his factory in Paris.

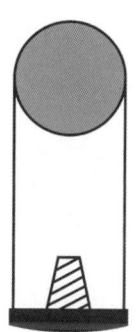

Pauly's patent of 1812.

In 1826 Casimir Lefaucheux, who had apprenticed as a gunsmith in his native district of Sarthe and later in Paris, was appointed as manager of the Pauly gun shop. For several years Casimir worked on improving Pauly's design, mainly in the effort to gain a gas tight seal between the barrel and the breech block. He gained his first patent, number 5525, in 1833 which improved the hinge which opened and closed the barrel. Cassimir Lefaucheux's fourth amendment to this patent included the drawing for the first, true, pinfire cartridge. The English translation for the patent reads: 'The base of my cartridge is drilled on the two opposite sides of its edge with round holes, which diameter is 9 points for one and 6 or 7 points for the other. The head of a pin is passed through the smallest of the holes and a copper percussion cap is pressed into the larger hole. The cartridge, thus armed, is loaded in the usual way, with only the end of the pin showing it offers the following shape. The base of the breech end of the barrel is slit in its upper part to accept the pin and it is struck in the usual way by a hammer of the piston type or any other spring system'.

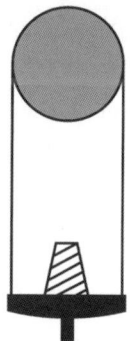

Pauly's modification of 1816.

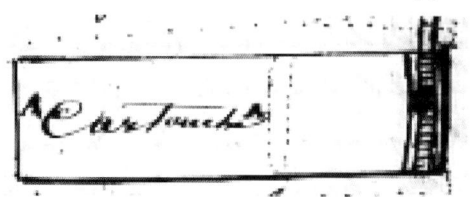

Plan and elevation of the Lefaucheux cartridge from his 4th addition to patent number 5525 of 31st March, 1835.

*Author's collection*

  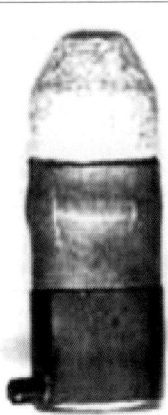

Early shotgun type pinfire cartridges. *left:* Lefaucheux type c.1840s. *centre*: Chaudun round c.1845. *right*: two part copper and paper cartridge c.1862.     *aaronnewcomer.com*

For many years the pinfire cartridge resembled a modern shotgun cartridge having a paper or card tube with a round or conical ball glued into the top, then in 1846 Houlier of Paris was granted a patent for an all metal cartridge which very quickly became the most popular type of cartridge for pistols.

Prior to the pinfire cartridge reloading a revolver required the shooter to press a paper cartridge into the chamber with an attached lever, then place a percussion cap on a nipple on the exterior of the cylinder. Now the simple action of placing the round into the rear of the cylinder loaded the weapon. The cartridges were also, for the most part, waterproof.

In the period 1847-71 nearly 60 pinfire cartridge patents were issued, some totally impractical others improving the system.

- 1847 Chaudun, cap containing primer made waterproof with application of resin.
- 1848 Beringer, lap jointed zinc case.
- 1850 Lefaucheux, copper case with separate screw on copper base.
- 1853 Gevelot, crown of paper between two copper cups to hold the primer.
- 1853 Boche, metal base with wound paper case.
- 1853 Devoir & Leclercq, conical base to powder chamber.
- 1854 Gevelot& Lemaire, metallic base and paper case.
- 1854 Chaudun, improved pin and wad.
- 1855 Vincent, metallic cap with concave powder chamber.
- 1855 Gevelot, paper tube with reinforced metallic base cap with paper wadding.

1855 Needham, tube and cap made of vulcanised rubber.
1855 Boche, Tordeux & Ouarnier, primer between pin in base and protruding pin.
1856 Perrin, shot and powder in separate compartments.
1857 Smith, cartridge constructed of vulcanised rubber.
1857 Houlier, end of tube rolled over card disc to retain shot.
1858 Thomas, ratchet teeth on cartridge to engage in a magazine.
1858 Ouarnier, screw in side of cartridge to act as retainer and anvil.
1859 Boche, wound paper partially conical powder chamber.
1859 Roy, paper tube with priming cap held within metallic lined enclosure.
1859 Chaleyer, pin passes completely through tube to assist in removal of spent cap.
1859 Varlet, pin passes through an external iron opening in the copper base cap.
1859 Lejeune & Chaumont, paper cartridge with steel, copper or brass base.
1859 Boche, bottle necked primer cup.
1859 Dougal, cartridge with conical tube to fire front end first.
1860 Chaleyer, perforated wad of non oxidizable metal, priming cup in perforation.
1860 Adams, shot and powder held separately in the case.
1860 Leme, shot and powder held separately in the case.
1860 Shedden, base wad covered by a metal disc which holds the primer.
1860 Rigby & Norman, vertical or horizontal pin driven flush with the case.
1860 Sear, case with hollow plug screwed into the base to give front ignition.
1861 Eley, means to prevent pin being blown when fired.
1862 Herbert, metal case with trapdoor in base which drops to expose priming cap.
1862 Sharp, thickening of the base to prevent bulging of the case on firing.

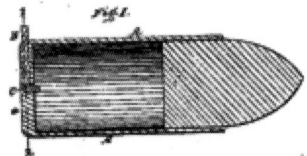

Sharps patent number 34,987 of 5th April, 1862.
*Author's collection*

1863 Brooman, pin shaped to prevent gas escape, paper case.
1863 Walsh, second pin on base to aid extraction.

1864 Wyley, fulminate placed between two pins one of which extends out of the case.
1864 Greener, primer fitted at the side of the case or at the end of a horizontal tube.
1864  Sneider, the case to separate on firing.
1864 Leetch, pin placed left of centre, rubber ring at top of tube to act as gas seal.
1864  Sneider, the case to separate on firing.

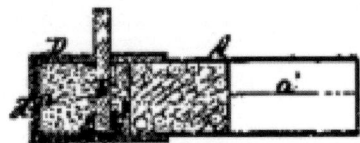

Snieder patent number 45210 of 22nd November, 1864.
*Author's collection*

1864  Leetch, pin placed left of centre, rubber ring at top of tube to act as gas seal.
1865  Thompson, pin forward in case, opening in side of case allows for capping.
1865  J & F Jones, methods of making cases of all types including pinfire.
1866  Clark (Gevelot), hole for pin made by punching, not drilling
1866  Sturtevant, cap on pin inserted from side, chamber is conical.

Sturtevant patent number 54038 of 27th March, 1866.
*Author's collection*

1866  Sturtevant, shoulder on pin acts as a valve to close the vent.
1866  Fitch, inside horizontal pin to the base of the case.

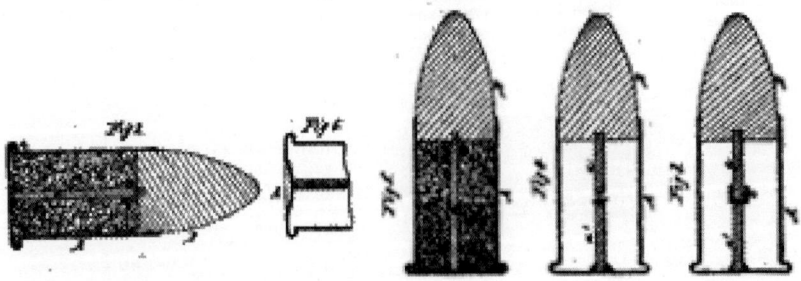

Fitch patent number 58800 of 16th October, 1866.  *Author's collection*

1867 Gedge (Audouy), iron or brass case with screw head to allow for re-priming.
1868 Rembert, pin extends across base to cap on opposite side
1869 Eley, placed a metal cup in front of paper base to prevent blow back of gasses.
1869 Wohlgemuth, two part cartridge with screw in base.

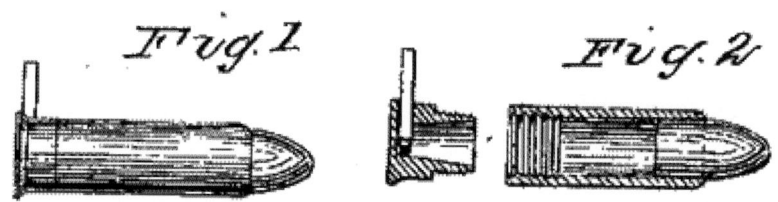

Wohlgemuths patent number 96373 of 2nd November, 1869.　　　Author's collection

1870 Lake (Sneider), case to separate on firing,
1870 Smith, radial opening in base closed by screw.

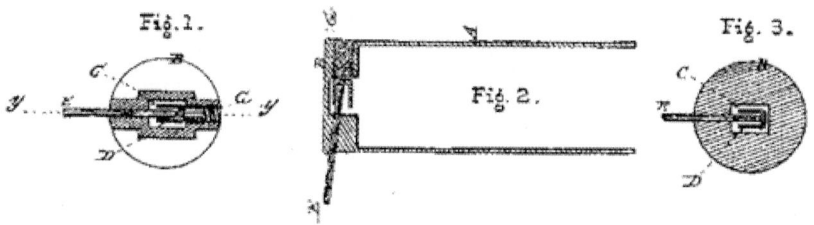

Smith patent 99721 dated 8th February, 1870.　　　Author's collection

1870 Gavard, rolled steel tube turned into a groove on the base plug.
1870 Sneider, flanged metallic base with pasteboard tube.

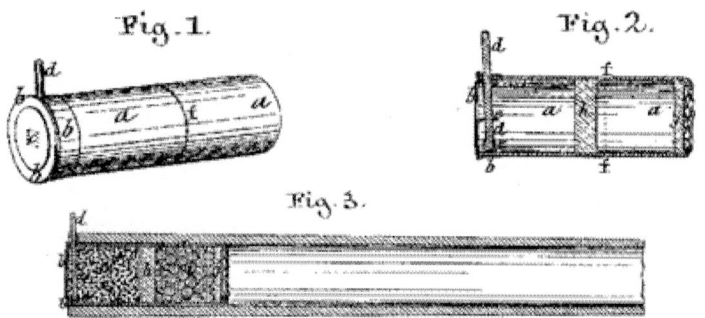

Sneiders patent number 102984 dated 10th May, 1870.　　　Author's collection

1870 Williams, pasteboard case with conical chamber and metallic base.

Williams patent dated 18th October, 1870, number 108543.    Author's collection

1871 Sneider, separate paper cups containing powder and shot.

Charles E. Sneider, Josias Pennington Jr., K. Nicholas G. Penniman.
Improvement in Shot-Cartridges.
116640                Fig. 1.              Patented Jul 4 1871

Sneider US patent 116640 of July 1871.    Author's collection

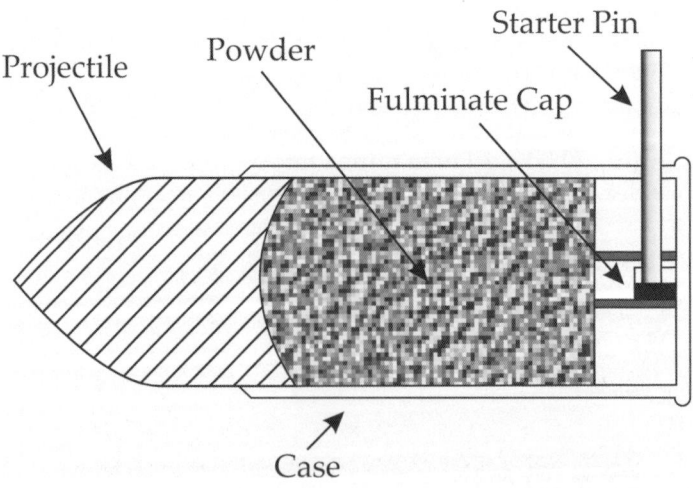

This diagram shows the pinfire cartridge as developed through the many patents issued in the 19th century. The hammer strikes the pin, forcing it onto the fulminate cap which ignites the main charge.    Author's collection

Given the overwhelming popularity of the pinfire system in Europe it is unsurprising that the vast majority of the cartridges were manufactured there, until the turn of the 20th century, when some manufacturers opened satellite factories in South America. The American continent had at various times only four pinfire cartridge manufacturers C D Leet, Sharps & Co, Allen & Wheelock and the Union Metallic Cartridge Co, the first named only producing for a few years during the Civil War.

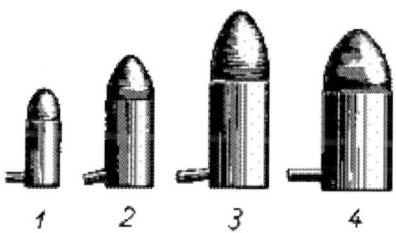

The standard pinfire calibres are 1. 5mm, 2. 7mm, 3. 12mm, 4. and 5. 15mm. 2mm ball or blank for miniature guns was also manufactured. The most popular calibres were 7. 9 and 12mm with 5 and 15mm being the least used. 15mm was, however extremely popular in South America with the Gauchos.

Lefaucheux-Rev.-Patrone
5 mm Kgl.

Lefaucheux-Rev.-Patrone
7 mm Kgl.

Lefaucheux-Rev.-Patrone
9 mm Kgl.

Lefaucheux-Rev.-Patrone
12 mm Kgl.

Lefaucheux-Rev.-Patrone
9 mm Schrot

Lefaucheux-Rev.-Patrone
9 mm Platz

Page from the 1900 Lindener Zundhutchen und Thonwaaren Fabrik of Linden, Hanover, catalogue of Georg Egestorff Clearly visible is the Anchor headstamp trade mark.

*Author's collection*

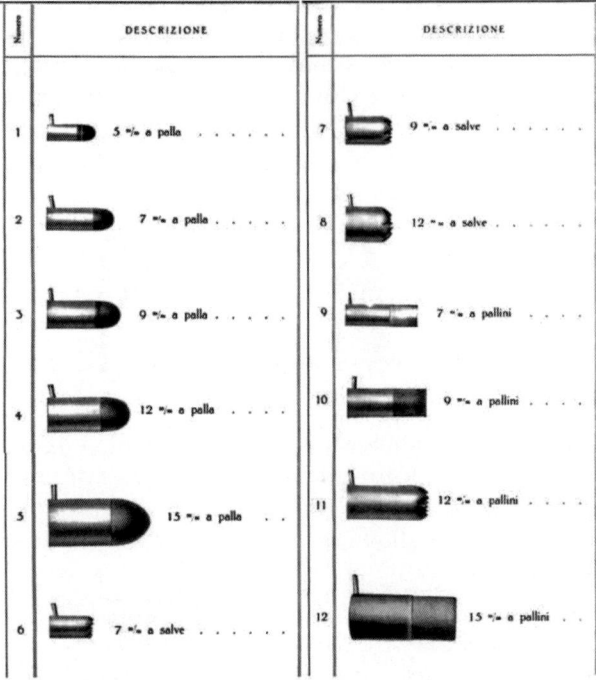

Pinfire page from the 1926 Fiocchi catalogue. These cartridges, manufactured in their plant in Lecco, Italy cover their complete range of ball, blank and shot. *Author's collection*

The perfect present for the pinfire shooter, a leather 20 round cartridge case by Manufacture Francaise d'Armes de Saint Etienne. *Author's collection*

# Chapter Five

# EUROPEAN PINFIRE AMMUNITION MANUFACTURERS

## AUSTRIA

### KELLER / HIRTENBERGER

Founded in 1860 by a German émigré Serafin Keller, the factory was located at Hirtenberg 24 miles southwest of Vienna.

In 1887 Ludwig Mandl purchased a half share and the company was renamed Keller & Co.

The Keller family sold the remainder of their shares to Ludwig Mandl in 1904 and it was renamed Hirtenberger Patronen Zundhutchen und Metalwarenfabrik AG, or Hirtenberger Cartridge Primers and Metal Goods Factory.

During the Second World War the factory was taken over by the Nazis to produce ammunition for the war effort. At its height over a million cartridges per day were being produced using forced labour from a sub-camp of the Mauthausen-Gusen concentration camp. At war's end the occupation forces stripped the factory and left it an empty shell.

In 1956 the ruin was purchased by the ex-manager who under the guidance of the previous owner, Fritz Mandl, re-started the munitions business as well as, from 1960, model aircraft engines.

*Above*: Three versions of the Keller & Co headstamp.
*Below*: Two Hirtenberger headstamps.

## BELGIUM

## LOUIS BACHMANN et Cie

Founded in 1858 Louis Bachmann et Cie was a small producer of ammunition in the Etterbeek district of Brussels. They ceased trading in 1889.

## A J DITS et Cie

Dits was an early supplier of pinfire ammunition from his factory in St Gilles, near Brussels. In 1862 he displayed a selection of pinfire cartridges at the International Exhibition in London. In 1865, he received an Honourable Mention at the Dublin International Exhibition.

## PIRLOT FRERES

Established in 1836 as an arms manufactory Pirlot Freres also sold ammunition under its own name. Known suppliers of pinfire ammunition to Pirlot were V Francotte May et Cie, Cartoucherie Belge and possibly Cartoucheries d'Anderlecht. Pirlot ceased production in 1879.

## VICTOR FRANCOTTE, and successors

1843 Louis Falisse set up as a percussion cap and ammunition producer
1863 The business was purchased by Messrs Wasseige and De Walque, the official title becoming Societe Wasseige, De Walque et cie, aka Capsularie Liege.
1867 The works moved to a more central position within Liege.
1871 Victor Francotte and his Partner Henry May purchase the business in 1871 and renamed it V. Francotte, May et cie but retained the trade name of Capsularie Liege.

1911 The company name was changed to Societe Anonyme Capsulalrie Liegeoise.

 Capsularie Liege

  Victor Francotte, May et Cie

Box of 50 7mm pinfire cartridges from the Liege factory of Victor Francotte, May & Cie.
*Author's collection*

## CHARLES FUSNOT, and successors.

1833 Charles Fusnot set up his percussion cap business in the Cureghem District of Brussels and within the next 20 years began the production of metallic cartridges including pinfire.

1882 The business was sold and the machinery was moved to the Anderlecht District of Brussels. The new title of the company was Societe Anonyme pour le Fabrication des Cartouches et Projectile.

1890 It was renamed as Societe Anonyme des Cartoucheries.

Box of 25 x 12mm ball cartridges by Societe Anonyme pour le Fabrication des Cartouches et Projectile.  *Author's collection*

1900 This Society, or grouping, was absorbed by Cartoucherie Russo—Belge which had offices in Brussels at 615 Rue St Leonard and Moscow in the Maryina Roshcha District.
1910 The group was once again renamed as Societe Anonyme Cartoucherie Belge,
1920 Fabrique Nationale of Herstal, Liege and Eley Brothers of London bought out SACB, the latter selling its share to F.N prior to the Second World War

Fabric de Balles et Cartouches Brussels

Societe Anoyme Cartoucherie Belgie, Brussels

Cartoucherie Russo-Belgie Brussels

Societe Anoyme Cartoucherie Belgie, Brussels

Societe Anonyme Cartoucherie Belge

**CHS. FUSNOT'S**
**CARTRIDGE CASES**
**NEVER MISS FIRE**
And are **THE BEST** and **CHEAPEST**.
Particulars of OSCAR SCHOLZIG, 9, New Broad St., London, E.C.

*Hartlepool Northern Daily Mail*, 10th January, 1884.   *N P Archives*

## ENGLAND

### ELEY BROTHERS

1828 William and Charles Eley established the company to produce patent cartridges.
1842 On the death of William his three sons inherited the business which was renamed Eley Bros.
1851 Eley Bros showed at the Great Exhibition at Crystal Palace.
1855 Eley Bros and Samual Colt registered a joint patent for revolver cartridges.

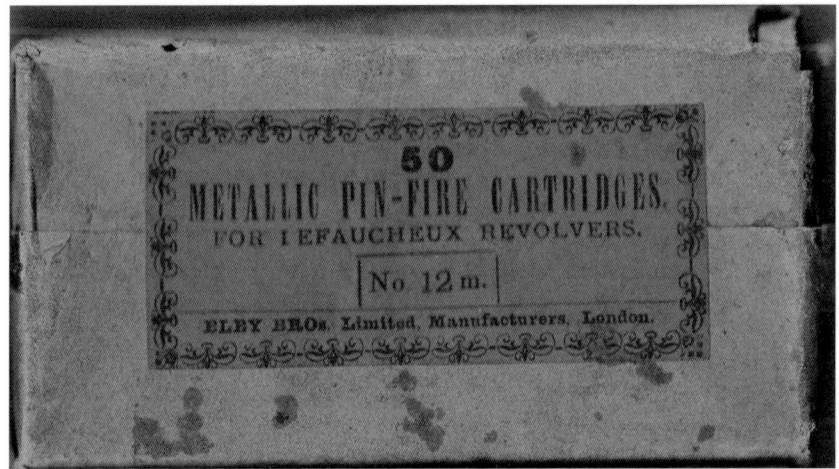

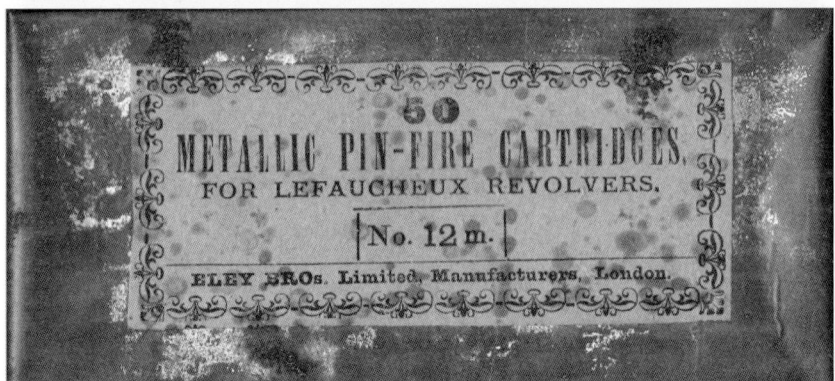

Early Ely Bros tin of 50 x 12mm pinfire cartridges with its cartridge paper outer package.
*Author's collection*

1857 Production began of Britains first centre fire ammunition.

1861-65 Ely produced millions of cartridges for the Confederate States of America during the American Civil War.

1874 Eley Bros went public with a shares issue which gave them an injection of much needed capital for a major expansion.

1895 An explosion at the Edmonton plant on 24th July killed five men. An official report blamed the mis-handling of a pinfire cartridge as the most likely cause.

1918 Following the end of the First World War much of the arms industry was in need of a cash injection to replace old, worn-out plant. The answer was to amalgamate several of the biggest ammunition producers into one company named Explosive Trades Ltd saving money on staff, equipment and factories. Those involved were Nobel Explosives, G Kynoch, Curtis & Harvey, Ely Bros, Kings Norton Metal Co and Birmingham Metal & Munitions Co.

1920 Explosive Trades Ltd was renamed Nobel Industries, but Eley Bros continued to trade under their own name.

1928 Cartridge production was moved to Witton in Birmingham and became a subsidiary of I.C.I Metals Division.

Cardboard box for 50 x 7mm cartridges. *Author's collection*

# European Pinfire Ammunition Manufacturers

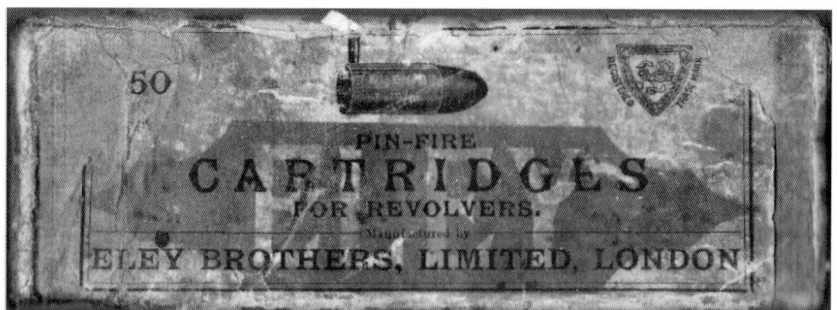

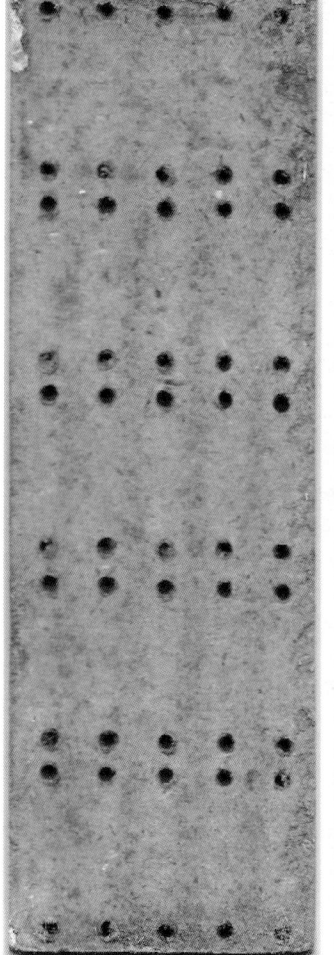

*Top*: Cardboard box of 50 7mm pinfire cartridges.

*Far left*: Internal laminated and drilled cardboard packing piece.

*Left*: Edge on view showing the top and tail method of packing each layer.

*Author's collection.*

## NATIONAL ARMS & AMMUNITION Co

THE NATIONAL ARMS and AMMUNITION COMPANY (Limited).—Incorporated under the Companies Acts, 1862 and 1867. Capital £300,000, in 15,000 shares of £20 each, of which the vendors take £100,000, in 5,000 paid up Shares, and 5,0000 Shares have already been applied for, leaving 5,00 Shares, which are now off--red for public subscription. £3 to be paid on application, £5 on allotment, £6 in three months, and £6 in six months. The whole or any portion of the capital may be paid up in advance of calls, and interest at £5 per cent. per annum will be allowed on the amount so paid in advance.

ABRIDGED PROSPECTUS.

The object for which this Company is formed are the establishment, upon a complete and extensive scale, of factories for the manufacture of breech-loading rifles and other arms, ammunition, projectiles, and war materials of all kinds, and for acquiring and working certain important patents and patent rights connected therewith, including the patents forming the system of Mr. Martini, which, after four years of exhaustive trials, has been adopted by her M-jesty's Government, in connection with Mr. Henry's rifle barrell, as the arm of the British service, under the designation of the Martini-Henry Rifle; also the sole right of manufacture of the systems of Mr. Peabody and Mr. Westley Richards.

Prospectus of the National Arms & Ammunition Company.  *Author's collection*

1871 The Westley-Richards Arms and Ammunition Co of Birmingham acquire the rights to the Martini-Henry rifle, the standard issue rifle of the British Army.

1872 A new company, the National Arms & Ammunition Co, was formed to manufacture the rifle and its associated ammunition as well as other small arms and cartridges. A small arms works was set up on Montgomery Street, Sparkbrook. The old Westley-Richards factory at Holdford Mills was used to manufacture metal cased ammunition.

1873 Cartridge works were located at Belmont Row and a general workshops set up at Peel Works on Macdonald Street which also manufactured bicycles and tricycles.

1874 To provide further capital for the ever expanding enterprise a Debenture Bond issue was made.

1878 The rapid production increase proved to be too much, too soon and the Belmont Row building had to be sold.

1882 Business was by now so bad that consideration was given to winding up the company and the government were approached as a possible buyer as this was where the British Army rifle was made.

1884 The bicycle manufactory was sold to a company from Coventry and the Peel Works was sold back to Westley-Richards. The government declined to buy the Sparkbrook and Holdford factories as they already had a Royal Small Arms Repairing Factory on Bagot Street.

National Arms 9mm  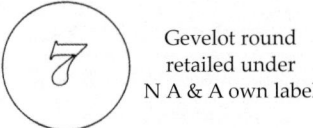 Gevelot round retailed under N A & A own label

1884 A change of heart meant the Sparkbrook Factory was eventually sold to the government, becoming Royal Small Arms Factory ( Birmingham )

1888 The final part of the NA&ACo, the Holdford Mill site was sold to the Gatling Gun Co.

## FRANK DYKE & Co

1883 Frank Dyke & Co was established on the upper floor of a warehouse on Dowgate Hill in the City of London. They were manufacturers of shotguns, wadding and shotgun cartridges. They also imported foreign weapons and ammunition for resale.

1893 By this date they had moved to 21 Addle Street, Aldermanbury, London, and later moved again to St Georges Avenue, Aldermanbury, London.

1907 On 31st December their premises were rocked by a huge explosion in the cellar which they used for cartridge filling. Four people were taken to hospital. All were quickly released when their wounds had been dressed, but 21 year old John Coker was found dead in the ruins. Exploding ammunition had sent balls across the street where they were buried up to half an inch into the walls. Possibly as a result of the explosion they were later recorded as having moved again, this time to 10 Union Street, London Bridge. Although they did not manufacture metal cased ammunition they did import it from Europe and relabelled and sold it wholesale to the gun trade.

*Lloyds Weekly* newspaper advertisement, 29th October, 1899.

National Newspaper Archive

Frank Dyke advertisements from 1927 (*Left*) and 1940 (*Right*).
*Grace's Guide to British Industry*

## KYNOCH & Co

1856   At the young age of 22 George Kynoch left his native Scotland to begin work at percussion cap manufacturers Pursall & Phillips of Whittall Street, Birminham.

1859   A massive explosion at the factory killed 19 of the 70 employees, including women and children.

1861   Pursall was granted permission to build a powder magazine and percussion cap factory on a four acre site at Witton, a rural hamlet three miles south east of Birmingham city centre.

The Lion Works pictured in 1867. The purchase in 1872 of a further four acres of land greatly increased the production area and many new workshops were built.

*Author's collection*

# European Pinfire Ammunition Manufacturers

**38     BIRMINGHAM ADVERTISEMENTS.     [1884.**

# G. KYNOCH & CO., LIMITED,
## WITTON, NEAR BIRMINGHAM,
# Ammunition Manufacturers.

AMMUNITION MANUFACTURERS BY SPECIAL APPOINTMENT TO H.M. THE KING OF SPAIN.

CONTRACTORS TO ALL THE PRINCIPAL GOVERNMENTS OF THE WORLD.

## SPORTING CARTRIDGES
### IN ALL BORES PAPER AND SOLID BRASS.

Gun Waddings, Percussion Caps, Anvils, and Re-loading Tools.—Punt Gun, Military, Express, Rook Rifle, and Revolver Cartridges of every kind.

## FOG SIGNALS.

## KYNOCH'S PATENT "PERFECT"
4, 8, 10, 12, 14, 16, 18, 20, 24, 28, ·410 & ·360 Bore, Pin Fire & Central Fire
## METALLIC CARTRIDGE.

"PERFECT" CASE, EMPTY.     "PERFECT" CASE, LOADED.

### THE CHEAPEST CARTRIDGE IN THE TRADE.
Attention is called to the New **PAPER-LINED "PERFECT."**
SAMPLES AND TESTIMONIALS FORWARDED ON APPLICATION TO THE WORKS.

**PRICE LIST AND ILLUSTRATED CATALOGUE TO THE TRADE ONLY.**

LONDON DEPOT:
**7 & 9, ST. BRIDE STREET, LUDGATE CIRCUS, E.C.**

BIRMINGHAM DEPÔT:—14, WHITTALL STREET.

An 1884 advertisement for Kynock & Co, Ltd following the management buyout.
*Grace's Guide to British Industry*

1862  George Kynoch became the proprietor of the company which was renamed Kynoch & Co.

1862/67  What started as two wooden sheds on the Witton site was by 1867 the large manufactory seen in the illustration on page 130. Named the Lion Works it had grown twenty fold in only five years.

1867/70  Kynoch made several advances in cartridge development which were protected by patents. Between 1868 and 1870 there were four explosions at the works. The worst of these was on 17th November, 1870 killing eight and injuring a further 20 persons.

1872  The size of the Lion Works was increased with the purchase of a further four acres of land.

1877  A rolling mill was leased on Water Street, Birmingham, for the production of brass sheet for metal-cased cartridges.

1884  George Kynoch was ousted as proprietor by a management buyout. However, he remained as managing director.

1886/89  With Kynoch as MD sharing his time between the works and his new job as MP for Aston quality and sales dropped, and costs rose. In 1888 Kynoch was forced to resign and a new MD was installed. The Water Street rolling mill, which belonged to Kynoch, was purchased and an option was taken to lease another mill on Lodge Road.

1891  Kynoch died in South Africa. The Water Street Mill was sold and all sheet metal work was transferred to Lodge Road.

1918  Kynoch & Co became part of a new engineering and armaments conglomerate. Initially named The Explosives Trade Ltd. It was very soon changed to Nobel Industries as they were by far the largest partner. The other companies included in the grouping were Eley Bros, Kings Norton Metal Co and Birmingham Metals and Munitions Co.

## CZECH

## SELLIER & BELLOT

1825 Louis Sellier, a German businessman and son of a French Royalist who fled his homeland during the Revolution, set up a percussion cap factory in Prague, Bohemia. He was quickly joined by his compatriot Jean Maria Nichlaus Bellot, another French émigré.
1829 A subsidiary plant was set up in Schonebeck, Prussia.

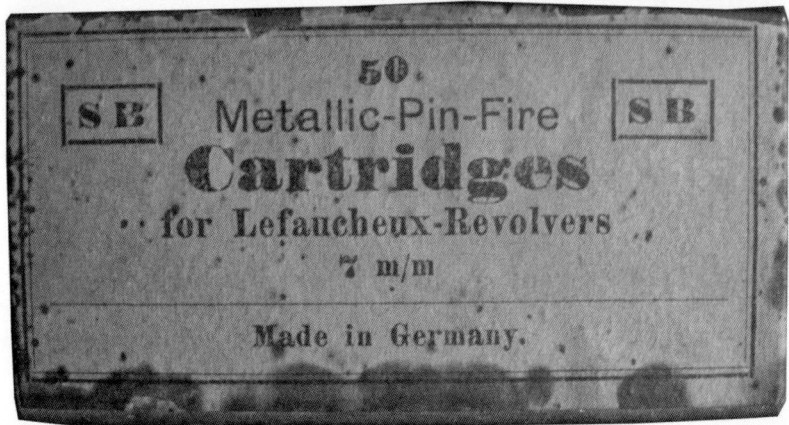

Two Sellier & Bellot tins for 50 x 7mm pinfire made in the Schonebeck factory.
*Author's collection*

1830 Between the two factories production reached over 60 million percussion caps per year.
1837 An exceptional year with over 156 million caps made.
1870 S&B began cartridge production with designs by Flobert (Rimfire), Lefaucheux (Pinfire) and the new Centerfire style. Production soon reached 10 million cartridges a year. Louis Sellier died and his heirs inherited his share.

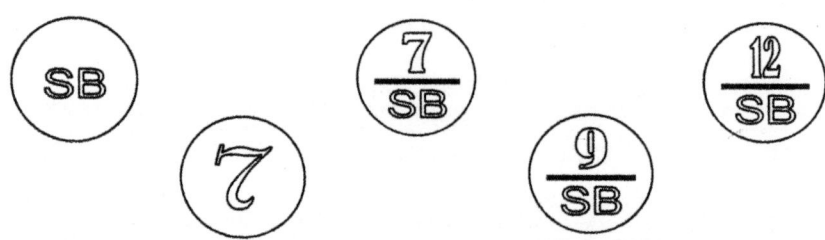

1872 An aging Jean Bellot was persuaded to give up running the factory and sell it. The new owner Martin Hala, a Czech entrepreneur quickly turned the business into a joint stock company.
1884 Another subsidiary factory was opened in Riga, Latvia which soon covered all demand for percussion caps in greater Russia and Scandinavia.
1893 The Sellier & Bellot Trademark was registered in Prague.
1895 Production begans of Hunting cartridges and copper primers for blasting works in a new plant at Skoda, Pilsen.
1914-18 For the duration of the First World War commercial ammunition production stopped and all efforts went into supplying the Army. After the foundation of the State of Czechoslovakia in October 1918 S&B become the main supplier to the Czech Army and Police.
1936 The company moved its production from Prague to a new factory in Vlasim 33 miles to the south east.
1939-45 Following the occupation of Prague by the Nazis on 15th March, 1939 all production was turned over to the German Army. At war's end in 1945 the new Soviet Regime S&B were nationalised on the introduction of a state monopoly for the manufacture of both military and commercial ammunition.

# FRANCE

## CARTOUCHERIE FRANÇAISE

1903  The company was established at Servilliers in the Val d'Oise. The two partners, Charles Gabel and Georges Leroy, set up in a small farm near the town with ten employees. The head office was in 8-10 Rue Bertin-Poiree, Paris. Gabel was a chemical engineer who had previously worked for the ammunition manufacturer Gevelot.
1910  With the business flourishing more workshops were built and more modern machinery installed.
1914-18  The Cartoucherie worked flat out to supply the French Army and now had over 2,000 workers, mostly women.
1918  As German advances moved the front nearer to the plant they are forced to temporarily relocate to the Military Pyrotechnics Factory in Caen.
1918-39  Work at Survilliers carried on as normal, returning to pre-war production of hunting and small arms ammunition.

Postcard of the entrance to the munitions factory in Vincennes, Paris, c.1900.

*Author's collection*

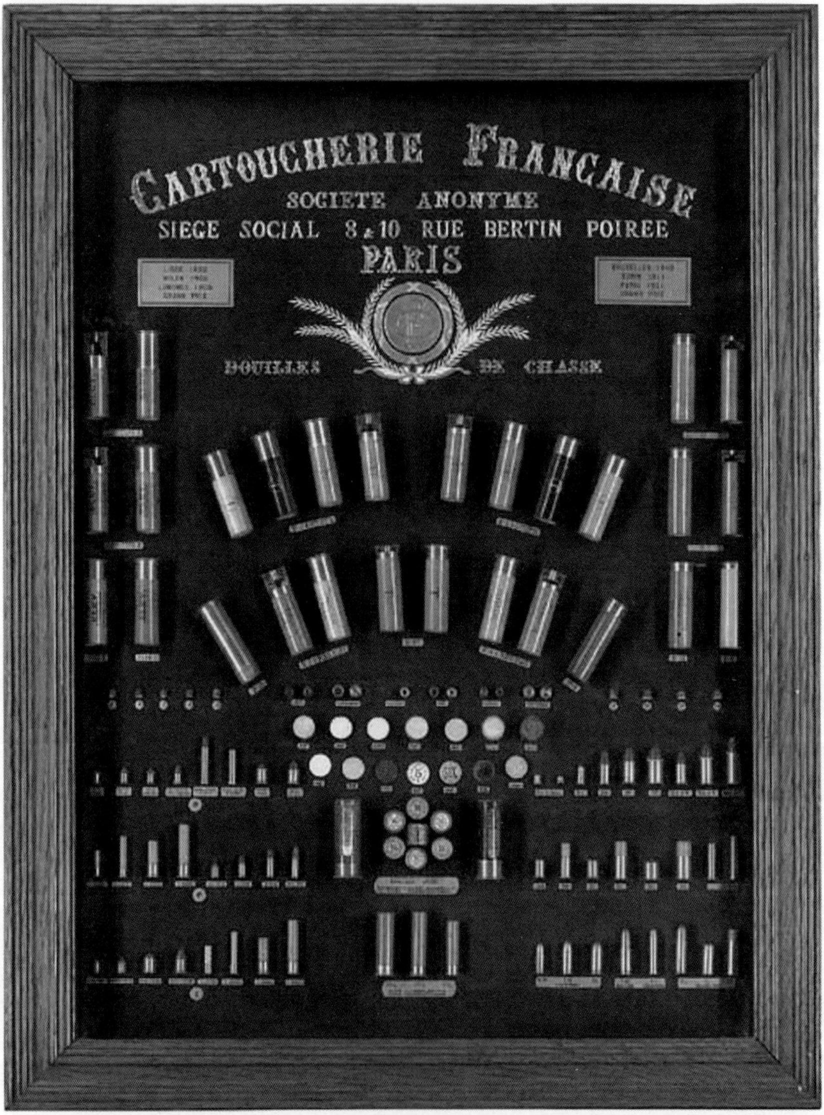

Display of cartridges produced by Cartoucherie Française.

Cartoucherie Française letterhead.

1939  The German invasion saw the works closed, with only a skeleton staff kept on to carry out small engineering works.
1945  Post-war production resumed, but never to the previous high volume.
1960-80  Orders slowly dried up with no military orders and the introduction of cheap foreign imports.
1985  A major fire on 16th and 17th January caused major damage to the works.
1989  All munitions work stopped and the factory began a new era as an explosive airbag inflator manufacturer.

**CHAUDUN et DERIVIERE**

Founded by Jules Joseph Chaudun in the first half of the 18th century this cartoucherie was amongst the first to produce pinfire ammunition. The address given for the company was 17 Rue du Faubourg in Montmartre, Paris. After the expiry of Lefaucheaux's patent in 1845 Chaudun began the manufacture of pinfire ammunition for Lefaucheaux marked with their Rue Vivienne address.

In 1847 Chaudun was granted a patent for his own improvement of pinfire cartridges. The main new ideas were to manufacture the case from copper or brass and to coat the primer with resin which rendered it waterproof.

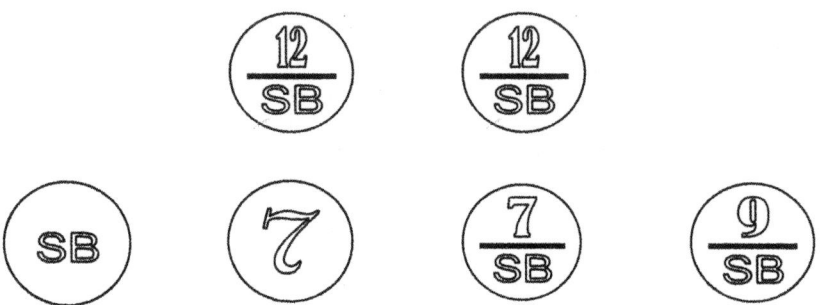

In 1860, following the death of Jules his son took over the business, and in December of that year a new partnership was formed with Monsieur N Deriviere.

## GEVELOT/SOCIETE FRANCAISE des MUNITIONS

Sometime in the first decade of the 19th century Joseph Marin Gevelot set up a business in Paris supplying Napoleons Grande Armie with knives and military equipment.

1816 Gevelot added arms and ammunition to the items supplied from his works on Rue Saint Denis
1820 Mass production of percussion caps for shotguns and pistols began.
1823 Joseph filed a patent for his invention of fulminate of mercury primers.
1826 Gevelot purchased the machinery and equipment from the defunct Leroy primer factory. To accommodate the increased manufacturing capacity a new works was built at Issy les Molineaux, Paris.
1835 A new head office was set up at 30 Rue Notre Dame des Victoires. By now the firm was manufacturing ammunition as well as primer caps.
1843 Joseph Gevelot died and left his widow Josephine and his 18 year old son Jules to run the business.
1853 On the death of Casimir Lefaucheaux, his 20 year old son Eugene took up the reigns at Maison Lefaucheaux. Childhood friends and neighbours, Jules and Eugene worked together to improve pinfire ammunition.
1867 The growing company now had over 500 employees and the green labelled cartridge boxes could be seen all over the world.

A 1920s style box of 9mm rounds.
*Author's collection*

1883 On the evening of Wednesday 28th February, at about 10 p.m the quiet of Paris was broken with the sound of a huge explosion. While moving barrels of black powder in the magazine a supervisor named Robert was blown to pieces when the powder accidentally ignited.

1884 Gevelot merged with another percussion cap manufacturer, Victor Gaupillat, and the company were renamed Societe Francaise des Munitions.

1888 On Friday 6th February a near miss occured at the London warehouse of Gevelot & Gaupillat on Upper Thames Street when the adjoining building was destroyed by fire. Luckily the explosives in the warehouse were not ignited. However the premises did incur some fire, smoke and water damage.

1898 The Issy les Molineaux seven hectare site now had over 50 buildings and exports account for half of the production.

1904 Jules Gevelot died and his widow, Emma, became president, a post she would hold for anther 23 years.

1973 A major fire destroyed half the Issy les Molineaux site, a blow from which the company never recovered.

1983 Cartridge production ceased and Gevelot became a car parts producer.

## HOULLIER & BLANCHARD

A high class Paris gunsmith whose premises were at 36 Rue de Clery. They also had a branch in Odessa, Russia from 1860-70. Their own-brand ammunition was manufactured for them by Gevelot.

Sealed box of 12mm cartridges by Gevelot for Houllier & Blanchard.

*sellantiquearms.com*

# GERMANY

## BRAUN & BLOEM

1848 Gustav Bloem opened a primer factory in a farmhouse at Derendorf, near Düsseldorf, with his friend Herr Nebe.
1849 An explosion destroyed the factory leaving Bloem penniless.
1850 Bloem entered into a partnership with Johann Heinrich Braun the wealthy owner of a metal fabrication works and they opened a primer factory in Ronsdorf. Braun & Bloem exhibited at the Expositition Universelle in Paris and won a medal.
1860 The company moved to Düsseldorf and began making Flobert and pinfire revolver ammunition.
1866 Bloem dissolved the partnership with Braun and became the sole owner of the company.
1905 Gustav Bloem died on 9th September and his family continued to own and run the works.
1918 The company was sold to Basse & Selve.

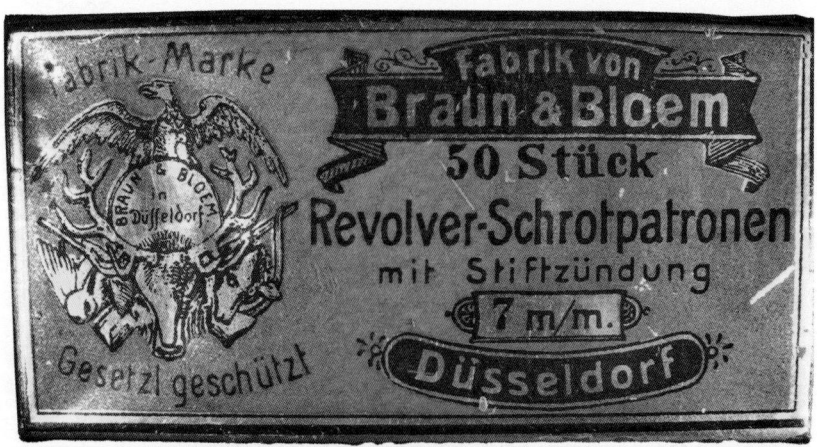

A 50 x 7mm shot cartridge tin for the home market.                    Author's collection

## GEORG EGESTORFF

1861 The Egestorff Zundhutchen Fabric was founded by Georg to manufacture primer caps, ammunition of all kinds and clay pigeons for the English market which were made from clay from the family brick works. By the mid 1860s the company was producing in excess of 300 million primers and rounds of ammunition a year.
1868 Georg died and the running of his empire was taken over by his family.
1879 The Egestorff empire was turned into a public limited company.
1913 In anticipation of impending military action and vastly increased production the munitions works were moved out of central Hanover to a greenfield site at Empelde.

The post-war depression and hyper-inflation was the death knell for the factory and it was sold to Dynamite Nobel who ran it until it was closed at the end of the Second World War.

Between 1831 and his death in 1868 Georg set up an empire of over 22 major industrial works including lime, salt, brick and chemical works, coal mines, paint factories and perhaps the largest of all the Hannoversche Maschinenbau Actien Gesellschaft—or as we know it now HANOMAG.

## GUSTAV GENSHOW & Co (GECO)

1887 25th April Genshow & Co opened as a weapons and ammunition wholesaler in Berlin.
1889 They acquired their first factory when they bought a bankrupt shot and ammunition works in Durlach.
1903 The company grew with the purchase of a hunting and sporting goods manufacturer in Hochenburg.
1906 Genshow opened a branch office in Cologne.
1924 The purchase of Deutsche Werke AG allowed for the manufacture of arms under their own trade mark of GECO (GEnshow CO).
1927 GECO entered into an agreement with Alfred Nobel & Co to share ammunition production, GECO made pistol ammunition for Nobel at its Durlach works and Nobel manufactured Flobert rounds for GECO.

Early Acorn headstamp

Post-1909 headstamp for ammunition made at the Durlach plant

1929   By this date GECO's three factories covered a total of 50,000 sq mtrs and employed 2,000 staff.
1938   Company interests in South America were in danger as there were many calls for nationalisation of arms and ammunition manufacturing. In an attempt to protect their investments GECO shares were transferred to influential Brazilian partners.
1940   Gustav Genshow died.
1943   All the company records were lost during an air raid on Hamburg in July.
1946   With the destruction of much of its plant during the Second World War GECO was unable to continue and was taken over by Dynamite Nobel.

## UTENDOERFFER / RHEINISCH-WESTFALISCHE SPRENGSTOFF

1856   Heinrich Utendoerffer set up a percussion cap factory in Nuremberg. Later they began manufacturing Flobert and pinfire ammunition.
1886   RWS founded in Cologne.
1889   Utendoerffer sold his business to the Rheinische-Westfalische Sprengstoff AG (RWS) who continued to retail Flobert ammunition under the Utendoerffer name.

1894   Unable to renew their licences in Nurenberg RWS closed their city centre factory and removed it to the more rural location of Stadeln, Fürth, where they continue to produce ammunition to this day.

Early tin of Utendoerffer 7mm pinfire cartridges.   *Author's collection*

# European Pinfire Ammunition Manufacturers 143

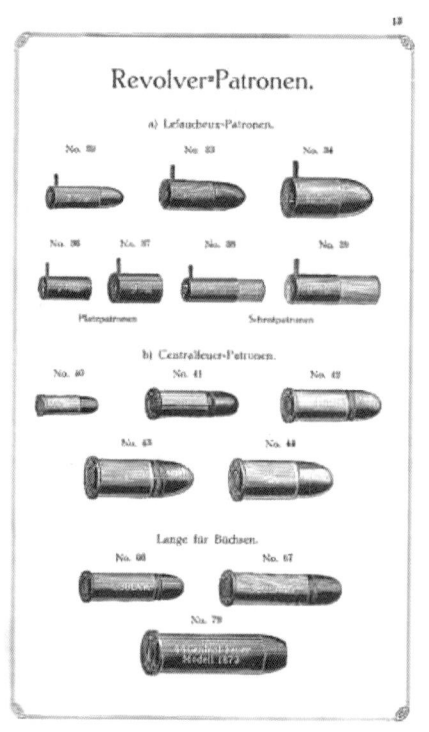

Front cover and pinfire page from the 1926 Utendoerfffer/RWS catalogue.
*Cornell Publications*

An RWS tin of 50 x 9mm blanks. *Author's collection*

## ITALY

### FIOCCHI

1876 Fiocchi Munizioni was founded by Giulio Fiocchi when he purchased the ammunition factory of a Lecco gunmaker.
1877 The new brand was launched selling primed cases for reloading

  fiocchi.com

1890s Fiocchi started to sell complete cartridges loaded with bullet or shot. At the same time they ceased making their own powder as it is proved to be cheaper to buy it in.
1930 Fiocchi began exporting ammunition around the world.
1945 The factory is destroyed by Allied bombing during the drive up Italy.
1946 Rebuilding of the plant was finished. Production continues to the present day.

The Fiocchi works at Lecco c.1900. *fiocchi.com*

## POLVERIFICIO PILONI

Owned by Bernardo Piloni the powder mill and ammunition works was opened in Bonacina, Lecco, in the last quarter of the 18th century. Built to the most modern design the risk of accidents should have been small. However, on 21st June, 1887 a major explosion destroyed the powder mill and killed seven workers.

After the death of Bernado the works were managed by his son Antonio and his nephews Bernado and Corrado. They specialised in the manufacture of gun powder, cartridges and general explosives as well as metal processing.

The powder and cartridge works were demolished in 2017

The Bernado Piloni Company. Award winning factory for common and fine hunting cartridges. Special Cartridges for compound nitro powders Accessories. National award winning powder factory for mine fuzes.
*resegoneonline.it*

Demolition of the Piloni powder and cartridge works in 2017.   *resegoneonline.it*

# SPAIN

## HERMANOS ARAMBURU / ARAMBURU BROTHERS

This was the Madrid Agency of Orbea Hermanos Fabrica. The proprietor was Joaquin Aramburu.

## ORBEA HERMANOS FABRICA / ORBEA BROTHERS WORKS

Founded in 1860 by the three Orbea brothers Juan, Mateo and Casimiro the company's main production was in small arms manufacture making handguns under licence for people such as Lefauchauex and Deprez. In the early 1880s they began producing cartridges for sale in Spain. In

The three Orbea brothers. *Orbea.com*

1897 the Spanish Government awarded a monopoly on gunpowder manufacturing to the Spanish Union of Explosives which seriously affected their production. To alleviate the problem in 1907 they opened a subsidiary company in Buenos Aires, Argentina which after two years was producing over 36 million cartridges per annum. Ironically most of these were exported to Spain.

Following massive overproduction of small arms during the First World War there was a depression in the arms manufacturing trade and so Orbea diversified into the manufacture of machine tools. By 1936 small arms production had ceased and the company had become the largest producer of bicycles in Spain.

c.1900 view of the Orbea Brothers factory in Eibar.    *Orbea.com*

## HIJOS de ORBEA FABRICA / SONS of ORBEA FACTORY

In 1927 the company split and some of the Orbea family moved to Vittoria, Alava to set up a cartridge manufacturing works. In 1946 the plant was moved to Iruna de laOca, Alava and renamed Explosivos Alaveses (EXPAL) where it would be absorbed by the Spanish Explosives Union.

## UNKNOWN MAKERS

Very plain German cardboard box of 25 x 7mm ball cartridges. Probably for small retailers, or wholesalers to put their own labels on.
*Author's collection*

Another German language carton, this time made of thick, sturdy cardboard. The large calibre, 15mm, and the heavy duty packaging implies they may be military issue.
*Author's collection*

This English language tin of 50 x 7mm pinfire cartridges is apparently nameless. However it is a standard Sellier and Bellot label, factory-trimmed to allow an English retailer to paste on their own name.
*Author's collection*

# Chapter Six

## UNITED STATES OF AMERICA PINFIRE AMMUNITION MANUFACTURERS

### ETHAN ALLEN / ALLEN & WHEELOCK

1831 Allen opened a cutlery and shoe factory in Milford, Massachusetts.

1832 The works moved to Grafton, Massachusetts. Making use of his metal working skills Allen began to make copper cased cartridges.

1836 Whilst working on a walking cane gun for a friend he designed an underhammer rifle and, with the help of his brother in law Charles Thurber, went into business as an arms maker. The firm was renamed Allen & Thurber.

1837 Once again the factory was moved, this time to a site in Norwich, Connecticut.

1847 The nomadic life continued with Allen moving to Worcester, Massachusetts.

1854 Thomas Wheelock, another brother-in-law, had been hired by the company a short time previously and in this year he was made a full partner. The company was renamed Allen, Thurber & Co.

1856 Charles Thurber died and the firm became Allen & Wheelock.

1862 With the American Civil War in full swing cartridges of all kinds were needed in huge numbers. The Union Army had purchased 20,000 Lefaucheux Model 1854 revolvers with ammunition and soon they needed to get more. Preferring to buy locally rather than import they put an order out to tender. Christian Sharps was given the contract for supplying the US army with 12mm pinfire ammunition, but supply was slow and the quality was appalling and so the order was re-let to William P Wilstach & Co of Philadelphia who sub-contracted the order to Allen & Wheelock. The initial order was for one million 12mm cartridges. The first batch of 202,800 were delivered to the Frankford Arsenal in May, and were accepted, but improvements in quality were asked for in the next delivery. In the following June and July the balance of the order was delivered without complaint. Allen approached the military authorities to see if he could supply the army direct but was informed that there would be no more pinfire orders as they now had all they needed.

An American Civil War period Allen & Wheelock cardboard pack of 24 12mm pinfire cartridges. At the top is the green labelled outer package, and below is the dark blue labelled inner.
*aaronnewcomer.com*

A post-Civil War box of 24 pinfire rounds sold after the death of Thomas Wheelock in 1865. The box has a deep blue label with silver text.
*aaronnewcomer.com*

1865 On the death of Thomas Wheelock, Allen took on his two son-in-laws, Sullivan Forehand and Henry Wadsworth. The firm was renamed Allen & Co.
1871 Ethan Allen died and the company was taken over by the son-in-laws and is renamed Forehand & Wadsworth. Metal cartridge making ceased and firearms continued to be the main product.
1890 Sullivan sold his shares to Forehand and the company is renamed the Forehand Arms Co.
1898 Sullivan Forehand died. His two sons run the company until 1902 when they sold to Hopkins & Allen who had been making their guns under contract.

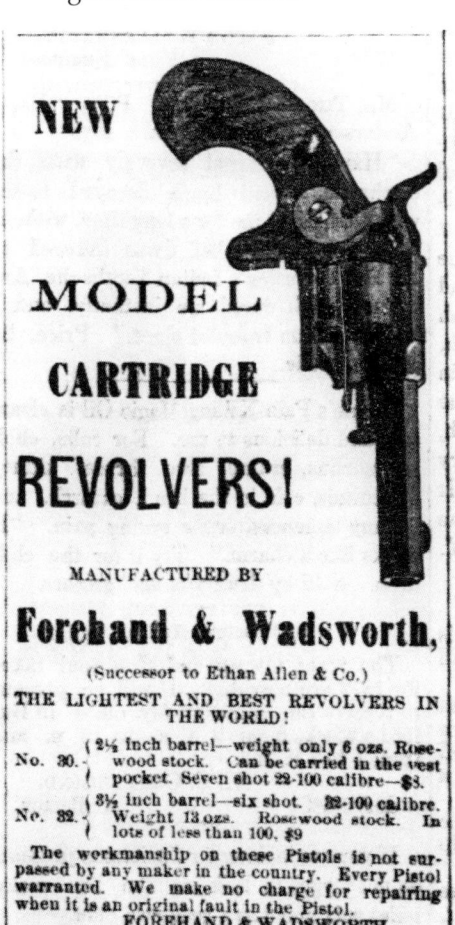

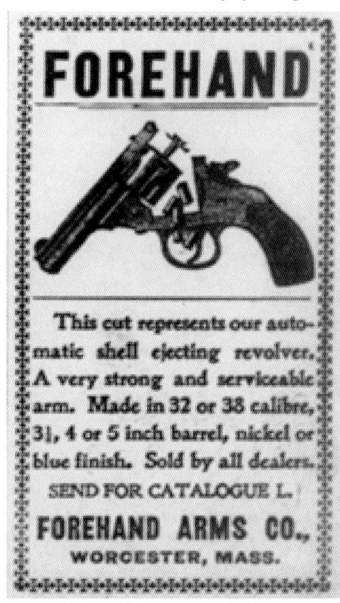

*Left:* An advertisement from the *Rock Island Daily Argos* for 17th December, 1872, not long after the death of Ethan Allen when the company became Forehand & Wadsworth.
*Library of Congress.*

*Above*: This advertisement from a February 1898 edition of the *Belle Plain News* shows the style of gun sold by the Forehand Arms Co.
*Library of Congress*

## C. D. LEET

1860 Charles Leet left his employment at Smith & Wesson to set up an ammunition works with a friend, Derrick Goff, titled Leet, Goff & Co. It was located on Market Street in Springfield, Massachusetts.

1862 Goff left the company and it was renamed C. D. Leet & Co. In February, to supply Union troops in the ongoing Civil War, the Frankford Arsenal order 250,000 12mm pinfire cartridges. On 14th April the first 50,000 were delivered quickly, followed in early May by another 88,000. The final 112,000 were delivered on 16th May. All deliveries were accepted without any problems over quality. In May another order was placed for a further 200,000 rounds which were delivered on 17th June.

1863 The last contract by the U S Government for any pinfire cartridges was awarded to Leet on 10th December when a further 75,000

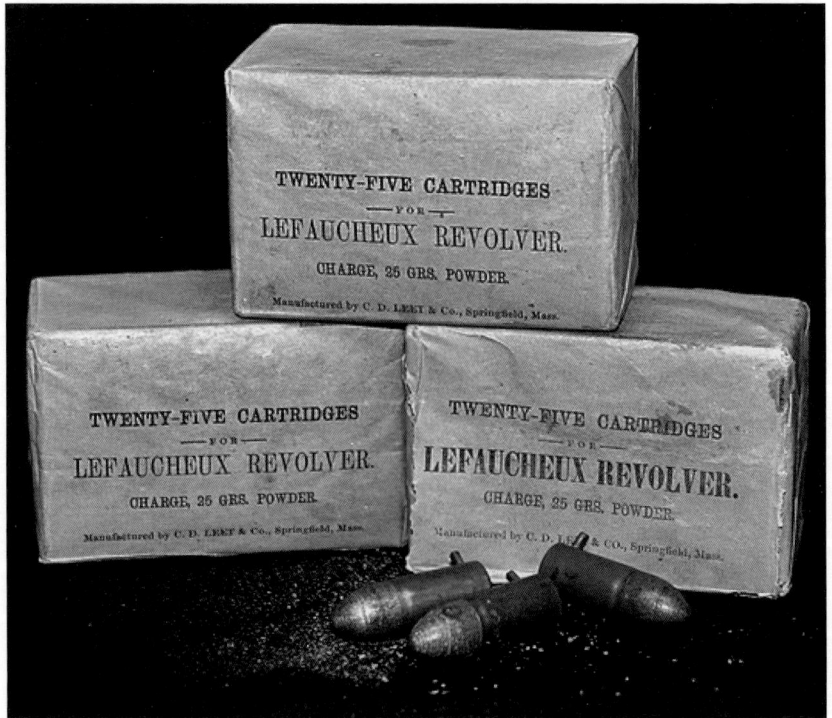

Three C. D. Leet & Co carton of 25 x 12mm pinfire cartridges in their brown paper wrappers as issued to the U S Army during the Civil War. Later packets were marked only C. D. Leet.
*aaronnewcomer.com*

were ordered. On 30th December Leet sent 76,000 to the Arsenal, accepted without query.
1864 The name of the firm is changed to C. D. Leet and all pinfire ammunition manufacturing machinery was sold to Schuyler, Hartley, Graham & Co, a New York sporting goods emporium.
1866 Charles' two sons joined the company. Leet continues to produce other cartridges in rim and centre fire configurations.
1876 Ammunition production ceased and the company was renamed The Rock Drill Company.

## C. SHARPS & COMPANY

1848 Christian Sharps began working in the gun trade at John Halls Harpers Ferry Gun Works. Whilst there he patented his first breech-loading rifle.
1850 He moved to Mill Creek, Pennsylvania where he had A. S. Nippes produce sporting rifles to his own design.
1851 A further move saw him in Hartford, Connecticut where he formed the Sharps Rifle Manufacturing Company. All of his weapons of this era were made under contract by Robbins & Lawrence of Windsor, Vermont.
1853 Sharps, who served as technical adviser to the company severed all links when the Sharps Rifle Manufacturing Co was declared bankrupt, though it continued to trade under his name. He moved again, this time to Philadelphia, and formed C. Sharps & Co where he continued to file rifle, revolver and ammunition patents.
1862 The imported French pinfire ammunition that came with the 12,281 Lefaucheux revolvers was found to be underpowered and prone to blowouts from the thin cartridge walls. In January the US military asked Sharps to provide one million 12mm pinfire cartridges to a new design that remedied those faults. The metal casing was thickened and made longer to hold a greater charge. He was contracted to have the first 50,000 cartridges at the Frankford Arsenal by 1st February, another 150,000 by the end of the month and a further 400,000 in March and April to complete the contract. By mid-February no cartridges had been delivered so the military put out feelers to other cartridge manufacturers to fill the gap in supply C. D. Leet were the first to benefit from Sharps' failings. On 20th March Sharps sent a letter explaining that the delay was due to problems with manufacturing the new design and that under Philadelphia law they could not have more than 25 pounds of

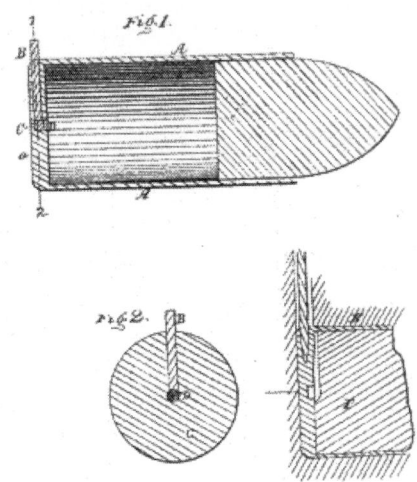

Christian Sharps' troublesome improved design for the pinfire cartridge contract for the US military during the Civil War.

*Author's collection*

gunpowder in the plant at any one time and therefore they had had to move the factory. Allen & Wheelock were given a contract to supply pinfire rounds to the US military. The 4th April saw the first test samples arrive at the arsenal, rejected as being useless. Sharps is joined by a new partner, William Hankins, and the firm renamed Sharps & Hankins. A further sample was received on 10th May and that too was rejected. The final delivery of 46,000 on 7th June was at last accepted but the contract was terminated and the only part of the order to be fulfilled.

## UNION METALLIC CARTRIDGE COMPANY (UMC)

1866 The large New York sporting goods and houseware emporium Schuyler, Hartley & Graham purchased two small New England

Two post-1873 Union Metallic Cartridge pinfire cartridge boxes, the 12mm had a blue label and the 9mm had a green label. UMC also produced 7mm pinfire rounds which had a blue label.

*aaronnewcomer.com*

## United States of America Pinfire Ammunition Manufacturers 155

ammunition manufacturers, C.D Leet of Springfield, Massachusetts, and Crittenden & Tibbals of South Coventry, Connecticut. The machinery of both was moved to a new facility in Bridgeport, Connecticut where ammunition production began under the name of the Union Metallic Cartridge & Cap Company.

1867 In September the Union Metallic Cartridge Company was incorporated.

1873 UMC began manufacturing pinfire ammunition in 7mm, 9mm and 12mm using the machinery last used in the Civil War by C. D. Leet.

1880 In September a huge explosion ripped through the fulminating building killing five members of staff, J. Sullivan, 15; J Tobin, 17; M Dempsey, 23; W Fuerchinger, 45 and P Clark, 50. The last two were both married with children and the newspapers carried graphic details of the dismembered bodies of two of them being pulled from a lake several hundred yards away.

1888 Schuyler, Hartley & Graham purchased the E Remington & Sons arms manufacturing company of Ilion, New York and rename it Remington Arms as the weapon arm of their sporting goods retail outlets.

1906 Another explosion rocked the Bridgeport plant as 16 tons of gunpowder exploded in a magazine on 14th May. This time there were no casualties, but damage was reported up to 30 miles away.

1909 A shot tower, 190 feet high was built in February. This was the tallest building in Connecticut for many years.

An advertisement from the *Orangeburg News* for the 2nd January, 1875 with Jas Brown offering UMC cartridges of all types.
*Library of Congress*

Graphic report of the 1880 explosion at the UMC works as described by a Kansas newspaper, the Iola Register.
*Library of Congress*

1912 Remington Arms and UMC merged and Bridgeport was named as the corporate headquarters while the arms side remained in Ilion, New York.

The Union Metallic Cartridge Company plant in Bridgeport, Connecticut on a 1917 postcard.
*Author's collection*

1933 DuPont purchased 60% of Remington-UMC.

1937 UMC ceased to exist as the company was wound up. Today the cartridges still carry the Remington UMC headstamp.

This advertisement from the *Philadelphia Evening Telegraph* of the 5th March, 1869 has 2,279 new, repaired and rusty mixed revolvers and 190,000 Lefaucheux and Wesson cartridges for sale at the Allegheny Arsenal, Pittsburg.
*Library of Congress*

**GOVERNMENT SALE.**

Will be sold at Public Auction, by H. B. SMITHSON, Auctioneer, at Allegheny Arsenal, Pittsburg, Pa., commencing at 10 o'clock A. M., Wednesday, March 24, 1869, the following articles, viz.:—

28 Cast Iron Cannon.
16,394 Solid Shot (round).
2,829 Stands of Grape and Carcasses.
3,827 Carbines, new, repaired, rusty, etc.
3,127 U. S. Rifles, Cal. 54 and 58, repaired, rusty, etc.
4,377 Enfield Muskets, repaired.
4,319 Foreign Muskets and Rifles, rusty, etc.
3,130 U. S. Muskets, Cal. 69, rusty, etc.
2,279 Pistols and Revolvers, new, repaired, and rusty.
4,000 Sets of Infantry Accoutrements (old).
33,182 Pounds of Cannon, Musket, and Rifle Powder.
190,000 Pistol Cartridges (Lefaucheux & Wesson's).
1,300,000 Maynard's and Sharp's Primers.
6,282 Musket Bayonets.
130,000 Pounds of Scrap Iron, Cast and Wrought.
A lot of Appendages and parts of Muskets.
A lot of Tools for Blacksmiths, Carpenters, etc. etc.
A lot of Packing Boxes, etc.
Catalogues of the above can be obtained on application to the undersigned.
Purchasers will be required to remove the property within ten days after the sale.
Terms—Cash.
R. H. K. WHITELEY,
2 22 mw 6t    Bvt. Brig.-Gen. U. S. A.

Union Metallic Cartridge Co cardboard counter advertisement dating from before the First World War. One of a series of posters which showed American wildlife.

*Author's collection*

# Appendix One

## OTHER EIBAR GUNSMITHS

Research into gunsmiths in Eibar has demonstrated a very large number of individuals who were involved in the trade there. This list greatly expands on the selected individuals that appear in the section on Spanish gunsmiths in Chapter Two on page 108.

**Aguirre Hermanos**, Juan and Martin. Gunsmiths, active 1883 to 1886. **Alberdi**, Domingo. Gunsmith, active 1870 until his death in 1888. **Alberdi**, Candide. Gunsmith, active 1881 to 1884. **Alberdi**, Felipe. Gunsmith, active 1886 to 1889. **Alberdi**, Juan Maria. Gunsmith, active 1883 to 1889. **Albizuri**, Maximo. Gunsmith, active in 1886. **Aldasoro**. Gunsmith, active 1860. **Aldazabel**, Jose Maria. Gunsmith, active 1884 to 1886. **Aldazabel**, Miguel. Gunsmith, active 1888 to 1894 and then again from 1914 to 1917. **Aldazabel y Zabala**. Gunsmiths, active 1887 to 1895. **Anitua**, Jose Francisco. Gunsmith, active 1883 to 1892. **Anitua y Charola**. Gunsmiths, active 1870 to 1898. **Anitua y Echiverria**. Gunsmiths, active 1884 to 1885. **Aramberri Hermanos**. Gunsmiths, active 1881 to 1884. **Aramberri**, Jose. Gunsmith, active 1883 to 1884. **Arana**, Esteban. Gunsmith, active 1883. **Arando y Azpiri**. Gunsmiths, active 1887 to1889. **Aranguren Familia**. Gunsmiths, founded in 1814 and active until 1912. **Aranzabal**, Hilario. Gunsmith, active c.1883. **Aranzabal**, Jose Antonio. Gunsmith, active 1883 to 1885. **Arguirro**, Nicolas. Gunsmith, active 1884. **Arguinao**, Gunsmith, active 1883. **Arino**, Thomas. Gunsmith, active 1883. **Aristondo**, Antonio. Gunsmith, active 1881. **Arizaga**, Francesco. Gunsmith, active 1881 to 1884. **Arizmendi**, Domingo. Gunsmith, active 1887 to 1914. **Arizmendi**, Everisto. Gunsmith, active 1884. **Arizmendi**, Jose Antonio. Gunsmith, active 1883 to 1889. **Arostegui**, Fernando. Gunsmith, active 1881 to 1889. **Arrate**, Daniel. Gunsmith, active 1883 to 1889. **Arrate**, Fernando. Gunsmith, active 1881 to 1889. **Arriola**, Jose. Gunsmith, active 1886. **Arrizabalaga**, Calixto. Gunsmith, active 1883 to 1889. **Arrizabalaga**, Juan Bautista. Gunsmith, active 1883 to 1889. **Arroyabe**, Santiago. Gunsmith, active 1883 to 1885. **Ascasibar**, Santiago. Gunsmith, active 1883. **Azpiri**, Domingo. Gunsmith, active 1883 to 1889. **Azpiri**, Juan Agustin. Gunsmith, active 1882 to 1894. **Azpitarte**, Manuel. Gunsmith, active 1884 to 1899. **Baroja**, Ventura. Gunsmith, active 1883 to 1886. **Barthelet**, Amando. Gunsmith, active c.1860 to 1880. **Bascaran**, Jose Maria. Gunsmith, active 1883 to 1890. **Bascaran**, Miguel. Gunsmith, active 1883 to 1884. **Bascaran**, Ronualdo. Gunsmith, active 1885 to 1889. **Basterrica**, Jose Antonio. Gunsmith, active 1883 to 1884. **Beristain y Hijo**. Gunsmiths, active 1881 to 1884. **Betolaza**, Alvaro. Gunsmith, active 1883 to 1894. **Bustinduy Cia**. Gunsmiths, active 1883 to 1885. **Crucelgui Hermanos**. Luis and Jose. Gunsmiths, active 1881 to 1894. **Ecenarra**, Anselmo. Gunsmith, active 1883. **Echaniz**, Cristobel. Gunsmith, active 1881 to 1896. **Echaniz**, Jose. Gunsmith, active 1883 to 1885. **Echeverria**, Jose Cruz. Gunsmith, active 1888 to 1894. **Echeverria y Hijos**. Gunsmiths, active 1886 to 1887. **Echeverria**, Pedro. Gunsmith, active 1883. **Eguiazu y Urizar**. Gunsmiths, active 1883. **Eguiguren**, Philip. Gunsmith, active 1881. **Elejalde**, Domingo. Gunsmith, active 1881 to 1883. **Elejalde**, Donato. Gunsmith, active 1881 to 1883. **Gabilondo-Urruzuno**, Jose. Gunsmith, active 1881 to 1888. **Garate**, Crispin. Gunsmith, active c.1883. **Garate**, Jerome. Gunsmith, active c.1883. **Garate, Arrananga y Cia**. Gunsmiths, active 1884. **Guenaga**, Vincent. Gunsmith, active 1886. **Guesalaga**, Jose Antonio. Gunsmith, active 1881 to 1884. **Guesalaga**, Angel. Gunsmith, active c.1860 producing pinfire revolvers. **Guesalaga**, Domingo. Gunsmith, active 1883 to 1885. **Guesalaga**, Florence. Gunsmith, active 1883. **Guisalaga**, Hilario. Gunsmith, active 1883. **Guisalaga**, Juan Maria. Gunsmith, active 1886 to 1889. **Guisalaga**, Pedro Martin.

## Other Eibar Gunsmiths

Gunsmith, active1883 to 1890. **Guisalaga Hijos de Pedro Martin.** Gunsmiths, active 1896 to 1899. **Guruceta,** Juan. Gunsmith, active 1887. **Irbarzabel,** Ignatius. Gunsmith, active 1881 to 1890. **Irbarzabel,** Juan. Gunsmith, active 1885 to 1889. **Irbarzabel,** Teodoro. Gunsmith, active 1881. **Inarra-Iraegui,** Juan Esteban. Gunsmith, active 1883 to 1888. **Inarra-Iraegui,** Pedro Maria. Gunsmith, active 1881. **Iraeta,** Jose Joaquin. Gunsmith, active 1883 to 1884. **Iraola,** Mateo. Gunsmith, active 1883. **Iraola,** Timoteo. Gunsmith, active 1886. **Iriondo,** Jose Martin. Gunsmith, active 1887 to 1892. **Iriondo,** Pedro. Gunsmith, active 1883 to 1892. **Iturbe,** Francisco. Gunsmith, active 1886. **Iturricastillo,** Martin. Gunsmith, active 1881 to 1893. **Iturrioz,** Antonio. Gunsmith, active 1883 to 1891. **Izaguirre,** Julian. Gunsmith, active 1885 to 1889. **Jarranaga,** Pedro. Gunsmith, active c.1860s. **Larranaga,** Juan Jose. Gunsmith, active 1881 to 1907. **Larranaga,** Pedro. Gunsmith, active 1883 to 1889. **Lizundia,** Antonio. Gunsmith, active 1881 to 1889. **Maitregean,** Felipe. Gunsmith, active 1885. **Maturana,** Felipe. Gunsmith, active 1883 to 1889. **Muguerza,** Marcos. Gunsmith, active 1881 to 1894. **Muguerza,** Pablo. Gunsmith, active 1881 to 1894. **Muguerza,** Pascual. Gunsmith, active 1883. **Ojanguren,** Agapito. Gunsmith, active 1883 to 1885. **Ojanguren,** Boniface. Gunsmith, active 1881. **Ojanguren,** Gabriel. Gunsmith, active 1885 to 1886. **Ojanguren,** M. Gunsmith, active c.1865. **Orozco,** Hermanos. Gunsmiths, Eibar Active 1881 to 1885. **Orozco y Sobrinos,** Gunsmiths, active 1886 to 1895. **Salaverria,** Blas. Gunsmith, active 1881 to 1892. **Salazar,** Jose Manuel. Gunsmith, active 1885. **Sarasqueta,** Ciriaco. Gunsmith, active 1883. **Sarasqueta Hermanos.** Gunsmiths, active 1887 to 1895. **Telleria,** Augustine. Gunsmith, active 1883. **Telleria,** Arsensio. Gunsmith, active 1886 to 1888. **Telleria,** Lorenzo. Gunsmith, active 1886 to 1888. **Trevino,** Blas. Gunsmith, active 1881 to 1889. **Ucin,** Manuel. Gunsmith, active 1883 to 1888. **Ugalde,** Joseph. Gunsmith, active 1883 to 1885. **Ugalde,** Vincenso. Gunsmith, active 1883 to 1884. **Unzueta,** Juan Esteban. Gunsmith, Guernica. Active 1885 to 1896. **Uria,** Ambrosio. Gunsmith, active 1881 to 1890. **Urigoen,** Miguel. Gunsmit, active 1883 to 1885. **Urigoen,** Manuel. Gunsmith, active 1886. **Urigoen,** Urquiola. Gunsmith, Eibar, Active 1883 to 1896. **Urigoen,** Pedro Maria. Gunsmith, active 1885 to 1886. **Vergara,** Basilo. Gunsmith, active 1885. **Ciriaco-Villar Hermanos.** Gunsmiths, active 1881 to 1889. **Zubiaurre,** Gregorio. Gunsmith, active 1883. **Zubizarreta,** Emilio. Gunsmith, active 1892. **Zulaica,** Marcelo. Gunsmith, active c.1860 to 1870. **Zulaica,** Narcissus. Gunsmith, active 1881 to 1890.

# References and Sources

## Books

Le Qui est Qui de L'Armourerie Liegeoise 1800-1950, by Guy Gadisseur and Michel Druart. Atlantica, 2005
Systeme Lefaucheux, by Chris C Curtis. Graphic Publications, 2002.
The Pinfire System, by Gene P Smith and Chris C Curtis. Bushman Bradshaw Publishing, 1980.

## Websites

| | |
|---|---|
| Aaronnewcomer.com | Specialist pinfire website |
| anexo:marcas_y_punzones_de_la_armeria_eibarresa | |
| | Very useful compendium of Spanish gunsmiths |
| armia-eibar.eus | Arms industry museum, Eibar |
| bertock.home.xs4all | Dutch pinfire website |
| britishnewspaperarchive.co.uk | UK newspaper archive |
| cartridgecollectors.com | Mine of information on pinfire ammunition |
| cartridge-corner.com | Ammunition identification resource |
| casimirlefaucheux.com | Pinfire forum |
| catalogacionarmas.com | Spanish arms listing |
| chrisferon.free.fr | Article on Paris gunsmiths |
| chroniclingamerica.loc.gov | Library of Congress newspaper archive |
| deisterbergbau.de | Eggerstorff archive |
| fiocchi.com | Fiocchi archive |
| fonderieguidoglisenti.it | Glisenti archive |
| freemycollection.com | Pinfire ammunition archive |
| germanhuntingguns.com | German arms manufacturers |
| gov.uk/search-for-patent | UK Patent Office search engine |
| gracesguide.co.uk | Graces Guide to British Industrial History |
| lefaucheuxnet.wordpress.com | Early pinfire resource |
| littlegun.be | Superb resource for Belgian gunsmith information |
| mallorquina.pagesperso-orange.fr/source/pageD.htm | |
| | Belgian gunsmith listing |
| newspapers.library.wales | Welsh newspaper archive |
| orbea.com/gb-en/about-us/1840-1929 | Orbea Brothers history |
| resegone Online - notizie da Lecco e provincia » La città che cambia volto: la storia del Polverificio Piloni | History of the Poloni powder plant |
| uspto.gov | U.S Patent Office search engine |
| wkc-solingen.de | Kirschbaum history |